Springer Theses

Recognizing Outstanding Ph.D. Research

Aims and Scope

The series "Springer Theses" brings together a selection of the very best Ph.D. theses from around the world and across the physical sciences. Nominated and endorsed by two recognized specialists, each published volume has been selected for its scientific excellence and the high impact of its contents for the pertinent field of research. For greater accessibility to non-specialists, the published versions include an extended introduction, as well as a foreword by the student's supervisor explaining the special relevance of the work for the field. As a whole, the series will provide a valuable resource both for newcomers to the research fields described, and for other scientists seeking detailed background information on special questions. Finally, it provides an accredited documentation of the valuable contributions made by today's younger generation of scientists.

Theses are accepted into the series by invited nomination only and must fulfill all of the following criteria

- They must be written in good English.
- The topic should fall within the confines of Chemistry, Physics, Earth Sciences, Engineering and related interdisciplinary fields such as Materials, Nanoscience, Chemical Engineering, Complex Systems and Biophysics.
- The work reported in the thesis must represent a significant scientific advance.
- If the thesis includes previously published material, permission to reproduce this must be gained from the respective copyright holder.
- They must have been examined and passed during the 12 months prior to nomination.
- Each thesis should include a foreword by the supervisor outlining the significance of its content.
- The theses should have a clearly defined structure including an introduction accessible to scientists not expert in that particular field.

More information about this series at http://www.springer.com/series/8790

Frederico Francisco

Trajectory Anomalies in Interplanetary Spacecraft

A Method for Determining Accelerations Due to Thermal Emissions and New Mission Proposals

Doctoral Thesis accepted by
the University of Lisbon, Portugal

Springer

Author
Dr. Frederico Francisco
Intituto Superior Técnico
University of Lisbon
Lisbon
Portugal

Supervisor
Prof. Orfeu Bertolami
Faculdade de Ciências
University of Porto
Porto
Portugal

ISSN 2190-5053 ISSN 2190-5061 (electronic)
Springer Theses
ISBN 978-3-319-38704-8 ISBN 978-3-319-18980-2 (eBook)
DOI 10.1007/978-3-319-18980-2

Springer Cham Heidelberg New York Dordrecht London

Softcover reprint of the hardcover 1st edition 2015

Printed on acid-free paper

Springer International Publishing AG Switzerland is part of Springer Science+Business Media (www.springer.com)

Supervisor's Foreword

More than forty years ago, on 2 March 1972, Pioneer 10 was launched from Cape Canaveral in a voyage to the far ends of the Solar System. About a year later, on 5 April 1973, Pioneer 11 followed suit. In their voyage Pioneer 10 would encounter Jupiter, while Pioneer 11 would also visit Saturn, delivering impressive pictures of these planets and their moon systems. The two spacecraft then followed hyperbolic orbits near the plane of the ecliptic to opposite sides of the Solar System. Initial projections on their lifespan ranged around seven years, after which the existing deep space tracking stations would no longer be able to determine their position and receive their data. However, further developments in tracking capabilities extended the probes' contact to an ever-increasing limit. Thus, although the Pioneer 10 mission officially ended on 31 March 1997, contact was kept until 7 February 2003, when it was at about 70 astronomical units from Earth, that is, around 100 thousand million kilometres away. A radio failure on 1 October 1995 and the increasing exhaustion of the power source rendered Pioneer 11 unable to contact Earth, when it was at approximately 40 AU from our planet.

After fulfilling their objectives so successfully, both missions were scaled down. In the course of events, tracking of the spacecraft became a routine training exercise for future space controllers. However, it was during this ordinary span of their voyage that Pioneer 10 and 11 rendered the riddle, which was until recently known as the Pioneer anomaly. Analysis of radiometric data from Pioneer 10/11 by NASA's Jet Propulsion Lab (JPL) team revealed the existence of an anomalous acceleration on both spacecraft, inbound to the Sun and with a presumably constant magnitude of $a_A \simeq (8.5 \pm 1.3) \times 10^{-10}\,\mathrm{m/s^2}$. This is not expected from the usual dynamics, which account solely for the gravitational pull and the outward solar wind pressure, both varying with the distance to the Sun r according to the inverse square law $1/r^2$. It was measured from 1980 on, when Pioneer 10 was at a distance of 20 AU from the Sun and the solar radiation pressure acceleration on Pioneer 10 had decreased to below $5 \times 10^{-8}\,\mathrm{m/s^2}$. This was possible because the Pioneer spacecraft were excellent for dynamical studies due to their spin-stabilisation and their great distances, requiring a minimum number of Earth-attitude reorientation manoeuvres for these deep space missions to go beyond the Solar System.

As a first attempt to explain this phenomenon, the JPL team, which unravelled the anomalous acceleration, has failed to properly account for thermal effects as a putative explanation for the anomalous acceleration. Indeed, thermal effects such as gas leaks, heat radiation due to the two nuclear generators placed on booms or to the instruments onboard would obey exponential decay laws and no combination of effects could result in a constant acceleration inbound the Sun.

The origin of this riddle has been the subject of heated debate since the late 1990s. If from one hand authors like Murphy, Katz and Scheffer defended that thermal effects could very well explain the anomaly, the plausibility of their arguments were not backed by sufficiently accurate calculations and could not account for a constant anomalous acceleration. Thus, alternative explanations ranged from basic engineering principles to unknown physics, such as new fields or new gravitational effects. From a theoretical point of view, this anomaly was puzzling indeed as one had to device a theory that breaks the usual law of gravity in a very subtle way, so to affect only the motion of the referred spacecraft, for there is no analogous effect for planets. Hence, one of the very foundations of General Relativity is at stake: the Weak Principle of Equivalence, which states that all bodies fall in a gravitational field at the same rate, independently of their mass or constitution.

If for a while it has been thought that an answer to the puzzle could be achieved only through a dedicated mission, it was realised by 2007 or so, that onboard thermal effects due to the plutonium sourced radiothermal generators (RTGs) and secondary heat generated by instrumentation could, as first claimed by Murphy, Katz and Scheffer and dismissed by the JPL team, be at the root of the anomaly and hence independent lines of enquiry were then pursued.

This thesis summarises the Ph.D. work developed by Dr. Frederico Francisco from March 2009 to December 2014 under my supervision and co-supervision by Dr. Paulo Gil from Instituto Superior Técnico. In this thesis the reader will have the privilege to read, first hand, the development of one of the approaches to compute the effect of onboard thermal sources. It starts with the development of the point-like sources method developed and detailed in Chap. 2. Actually, as described in the text, the method was developed and thoroughly scrutinised by a series of test cases. The procedure was completed with a brilliant touch, entirely due to Dr. Frederico Francisco, who realised that using a computer graphic technique, known as Phong shading, would allow for computing the specular reflection, whose suitable accounting yields the remaining 30 % or so of the anomaly that was still lacking in the previous calculations. Another extremely relevant contribution found in this work is the demonstration that one could advantageously consider a Monte Carlo analysis of the parameters to overcome the lack of knowledge of their values after more than thirty years of wandering in space. This analysis allows for the claim that the Pioneer anomalous acceleration can be fully accounted for by onboard thermal effects.

The results presented in this work preceded the ones presented by ZARM's group in Bremen and the one by the JPL group, in particular, by over a year. These groups used essentially the well-known finite element approach.

The method developed in Chap. 2 is then used in Chap. 3 to compute the acceleration due to thermal effects originated by the RTGs on board the Cassini spacecraft. Remarkably, the obtained values match extremely well the observed ones obtained by Doppler shift measurements and strongly support the most stringent test of General Relativity through an estimate of the γ parameter of the Parameterised Post-Newtonian formalism.

In Chap. 4, a description of the Outer Solar System mission proposal is presented, where to the Portuguese team, of which Dr. Frederico Francisco is a member, has participated in the design of its scientific goals based on previous work on the gravitational signature of the Kuiper Belt.

Finally, in Chap. 5, the reader is introduced to yet another anomaly, the flyby anomaly, a discrepancy between the incoming and outgoing velocities of spacecraft that perform gravity assists nearby Earth. If at all real, the origin of this effect remains elusive, in Chap. 5 of this thesis, a quite interesting proposal to better characterise and analyse this anomaly is presented and that uses the Galileo Navigational Satellite System. It also includes a quite interesting mission proposal to test the existence of this anomaly in highly eccentric elliptic orbits.

Finally, I would like to thank Frederico Francisco for being such a splendid Ph.D. student, and the other members of the Portuguese research team, Drs. Paulo Gil and Jorge Páramos for the work developed over the years. Thankful thoughts are also due to the various colleagues with whom we have worked on the Pioneer anomaly and other space science problems. A partial list of their names includes Clovis de Matos, Slava Turyshev, Serge Reynaud, Hansjorg Ditus, Klaus Lämmerzahl, John Moffat, Bruno Christophe, Peter Wolf, Viktor Toth, Craig Markwardt, John Anderson, Pierre Touboul, Robert Bingham, Ulrich Johann, Pierre Bouye, Lucianno Iess and Wolfgang Ertmer.

I would like also to express my deep gratitude to Angela Lahee and to Springer Verlage for the far-sightedness of generously publishing this thesis.

Porto
March 2015

Prof. Orfeu Bertolami

Abstract

The so-called Pioneer anomaly, an anomalous acceleration that was detected in these missions, stood out for many years as a potential deviation from General Relativity. It sparked many explanation attempts and generated great controversy around the true value of thermally induced acceleration.

A new method was developed to address this issue. It is presented here along with the battery of tests performed to ensure the reliability of its results. The method is based on a distribution of point-like sources and a reflection model inspired by computer graphics techniques.

The results show that the Pioneer anomaly can be fully explained through anisotropic thermal radiation originating from the probes themselves.

The same method was also used to model the non-gravitational acceleration detected during a gravitational experiment performed in the Cassini mission. It is shown that the measured acceleration can be modelled by thermal effects.

The Pioneer and Cassini gravitational tests inspire a proposal of a new outer solar system mission that can improve on them.

One outstanding trajectory anomaly is the so-called flyby anomaly. Due to the absence of a full characterisation, this thesis proposes a new experimental method to study this effect.

Acknowledgments

During the time it took to carry this doctoral degree to its conclusion, the number of people who, somehow, contributed to its successful outcome would be too large to name here. Still, it is important to highlight a few people who were more closely connected with this enterprise:

To my supervisors, Orfeu Bertolami and Paulo Gil, for the opportunity they have given me to work in these very interesting subjects in an intellectually stimulating environment and for the confidence they gave me when mine faltered.

To Jorge Páramos, for his friendship and for the valuable insight he often gave me, not only on scientific issues, but also on the personal and social aspects of research.

To Ana Margarida, for the love and patience she always had for me when I needed.

Finally, to all the friends and family that gave me support, that asked how it was going, that presented me with opportunities to be involved in other activities besides the academic work, that made these last four and a half years a great time to remember.

The work leading to this Doctoral thesis was sponsored by the FCT—Fundação para a Ciência e Tecnologia, under the grant SFRH/BD/66189/2009.

Contents

About the Author

Frederico Francisco obtained his Master of Science degree in Aerospace Engineering at Instituto Superior Técnico (IST), Lisbon, in 2009 with a dissertation containing an analysis of the Pioneer Anomaly, including a preliminary outline of thermal effects involved.

He then continued to work on this subject in his Doctoral thesis, presented in this book, with the development of the method that ultimately was able to solve that famous anomaly. The thesis also included an analysis of a gravitational experiment performed by Cassini, and the proposal of new methods for the next generation of tests on gravity in the solar system. He obtained his Ph.D. in Physics at IST, Lisbon, in December 2014 under the supervision of Prof. Orfeu Bertolami and Prof. Paulo J.S. Gil.

Besides the main research focus in space physics and solar system tests on gravitation, Frederico has also maintained collaboration with a research group at IST working on rail infrastructure maintenance and management.

During his time as a student, Frederico Francisco was deeply involved in university politics, serving two terms as Vice-president of the Pedagogical Council of IST, between 2009 and 2012, and one term as member of the General Council of the University of Lisbon, from 2013 to 2014.

Acronyms

ESA	European Space Agency
GNSS	Global Navigation Satellite Systems
GEO	Geosynchronous Earth Orbit
GR	General Relativity
HEO	Highly Elliptical Orbit
JPL	Jet Propulsion Laboratory
KBO	Kuiper Belt Object
MEO	Medium Earth Orbit
MLI	Multi Layer Insulation
NASA	National Aeronautics and Space Administration
PPN	Parameterised Post Newtonian
RTG	Radioisotope Thermal Generator

Chapter 1
Introduction

Since the dawn of experimental gravitational physics, the solar system has been one of its key testing grounds. The observation of anomalies in the trajectory of solar system bodies relative to theoretical prediction has often led to theoretical breakthroughs. Perhaps the best known example is the perihelion precession of Mercury. At first attributed to systematic errors, then turned into a major problem for Newtonian gravity and became one of the key experimental evidence supporting General Relativity (GR).

Despite its well known shortcomings, GR is still the best available theory of gravity. It has successfully resisted ever more sophisticated experiments and ever more accurate measurements made over the years [1]. In November 2015, less than a year from the moment this thesis is being written, the theory will complete its centenary.

As humanity started to launch its first interplanetary space probes in the 1960s, a whole new opportunity opened up for the testing of GR. The radiometric techniques developed to track these probes along their long trajectories allowed for very precise kinematic measurements.

The 1970s saw the first spacecraft being launched towards the outer solar system, beginning with Pioneer 10 and 11, respectively, the first human spacecraft to visit Jupiter and Saturn. These particular probes had some features that made them especially suitable to test theories of gravity. Indeed, the fact that they were spin-stabilised, requiring very few manoeuvres along their long trajectories, put them among the spacecraft with the most precise trajectory determination ever.

It turns out that the tracking of the Pioneer 10 and 11 deep-space probes led, as revealed in 1998, to the detection of a deviation from the predicted trajectory that could be translated into an additional anomalous acceleration acting on the spacecraft [2]. The discovery and subsequent confirmation of this acceleration, that became known as the *Pioneer anomaly*, generated a considerable enthusiasm among gravitational physicists, as it raised the possibility of new physics beyond GR. The reader can find a detailed discussion on the tracking methods used to detect the Pioneer anomaly in Ref. [3] and an extensive review of the attempts made to explain the Pioneer anomaly through new physics in Ref. [4].

F. Francisco, *Trajectory Anomalies in Interplanetary Spacecraft*, Springer Theses,
DOI 10.1007/978-3-319-18980-2_1

The Pioneer anomaly remained controversial for quite some time, as the discussion was focused on a constant acceleration, as determined by the initial analysis. This was incompatible with a thermal explanation, a possibility that had been raised since the beginning, but was met with some resistance from some members of the team that originally put forward the anomaly [3]. Subsequent independent data analyses showed that the data was actually consistent with an exponentially decaying acceleration, as would be the case if thermal effects were responsible [5].

The emergence of this problem raised, for the first time, the need for an accurate prediction of the acceleration induced by onboard thermal emissions. It was thus deemed that conventional methods would not be the best suited to solve this problem. One of the main outcomes of this thesis is the development of a whole new approach. A method was thus developed to tackle the specific challenges of modelling thermal accelerations in spacecraft with limited information. The method and its preliminary results, that already showed that thermal effects were a strong possibility to explain the Pioneer anomaly, were initially met with some skepticism, mainly from some JPL scientists that were involved in the study of this problem. This also meant that our final results would endure a longer than usual process leading to their publication, facing an intense scrutiny.

In the end, it was definitely shown that the Pioneer anomaly could indeed be explained by the effects of thermal radiation being emitted anisotropically from the probes themselves. Our results were subsequently confirmed by the other two teams working out this issue independently.

The development of the point-like source method and the process of obtaining the solution is explored in detail in Chap. 2, documenting the results published in the literature during the development of this thesis [6, 7], which complete the narrative started in this author's Master thesis and in the previously published preliminary results [8].

The physics involved in the thermal analysis method that was developed is fairly straightforward, relying mainly on classical electromagnetism, specifically, on the definition of the Poynting vector as the energy flow per unit area and the notion of radiation pressure. In order to grasp the full details of the formulation, the reader should also be comfortable with vector field integration. Any good undergraduate textbooks on electromagnetism and multivariable calculus should provide the necessary clarification for any doubts that should arise when reading the description of the method.

Some more recent interplanetary missions were designed from the beginning to perform experiments on GR. This was the case of Cassini's radiometric experiment, which made the most accurate measurement of the γ parameter of the Parameterised Post-Newtonian (PPN) formalism ever, and extended the almost century long run of GR experimental confirmations [9]. The PPN formalism is a formulation of Einstein's equations in terms of their lowest order deviation from Newtonian gravity. It provides an extremely useful set of parameters that can be tested with the appropriate experimental setups. A good review on the PPN formalism, its parameters and the experimental configurations used to test them can be found in Ref. [10]. A more detailed discussion about the Cassini solar conjunction experiment is carried out in Ref. [11].

To ensure the accuracy of the measurement, it is essential to carefully model any systematic non-gravitational effects that could induce errors in the tracking of the spacecraft's trajectory. While there are radiometric techniques that allow these effects to be singled out, they cannot necessarily be associated with a specific cause. For example, one can find the acceleration component that is constant relative to the spacecraft's attitude and associate it with its thermal emissions. However, these measurements often involve large errors, as well as the assumption that no other effects are present. The only way to achieve reliability in this particular situation is to obtain the acceleration estimate from a model of the spacecraft and compare it with the radiometric measurements.

This is precisely what was done, resorting to the same approach developed to study the Pioneer problem. This issue is further developed in Chap. 3, where it is shown that the result from the thermal modelling is compatible with the radiometric measurement and, consequently, validates the measurements from the gravitation experiment in question, as also published in Ref. [12].

The examples of Pioneer and Cassini as test beds for GR, pave the way for the next generation of gravitational experiments in the solar system. Chapter 4 briefly addresses the Outer Solar System (OSS) mission proposal. Its scientific objectives include experiments similar to those in the two preceding cases, but taking the next incremental steps to increase their accuracy. This proposal led to a publication for which this author contributed in the context of his doctoral work [13].

Many of these interplanetary probes used gravity assists to reach their targets with less fuel. Also in the last two decades, some questions have arisen about a velocity shift near the perigee of the hyperbolic trajectories of spacecraft near Earth. By analogy with the Pioneer anomaly, this became known as the *flyby anomaly*. Interestingly, both "anomalies" were presented in the peer-reviewed literature in papers that have John D. Anderson, a NASA veteran, as first author [2, 14].

This problem has been outstanding without any convincing explanation or even a coherent characterisation of the phenomena involved. Still, even with the scarce information available, attempts to consider some gravitational solutions have been put forward. The reader should find background on the flyby anomaly mainly in Refs. [14, 15], although he will find that the information is fragmented and sometimes inconsistent.

In Chap. 5, the available information on the flyby anomaly is reviewed and its inconsistencies scrutinised. Some of the proposed explanations are also briefly mentioned. The ultimate goal of the discussion around the flyby anomaly is to present a new experimental method. Given the lack of a workable characterisation of the anomaly, the emergence of more accurate spacecraft tracking solutions using Global Navigation Satellite Systems (GNSS) presents an opportunity to build low cost missions that could study the dynamics of spacecraft in orbits similar to those involved in the anomalous flybys. This mission concept was presented to the community in Ref. [16].

References

1. O. Bertolami, J. Páramos, *The Experimental Status of Special and General Relativity*. Springer Handbook of Spacetime (Springer, Berlin, 2014), pp. 463–483
2. J. Anderson, P. Laing, E. Lau, A. Liu, M. Nieto, S.G. Turyshev, Indication, from Pioneer 10/11, Galileo, and Ulysses data, of an apparent anomalous, weak, long-range acceleration. Phys. Rev. Lett. **81**(14), 2858–2861 (1998). doi:10.1103/PhysRevLett.81.2858
3. J. Anderson, P. Laing, E. Lau, A. Liu, M. Nieto, S.G. Turyshev, Study of the anomalous acceleration of Pioneer 10 and 11. Phys. Rev. D **65**(8), 082004 (2002). doi:10.1103/PhysRevD.65.082004
4. S.G. Turyshev, V.T. Toth, The Pioneer anomaly. Living Rev. Relativ. **13**, 4 (2010). doi:10.12942/lrr-2010-4
5. S.G. Turyshev, V.T. Toth, J. Ellis, C.B. Markwardt, Support for temporally varying behavior of the Pioneer anomaly from the extended Pioneer 10 and 11 doppler data sets. Phys. Rev. Lett. **107**(8) (2011). doi:10.1103/PhysRevLett.107.081103
6. O. Bertolami, F. Francisco, P.J.S. Gil, J. Páramos, Estimating radiative momentum transfer through a thermal analysis of the Pioneer anomaly. Space Sci. Rev. **151**(1–3), 75–91 (2010). doi:10.1007/s11214-009-9589-3
7. F. Francisco, O. Bertolami, P.J.S. Gil, J. Páramos, Modelling the reflective thermal contribution to the acceleration of the Pioneer spacecraft. Phys. Lett. B **711**(5), 337–346 (2012). doi:10.1016/j.physletb.2012.04.034
8. O. Bertolami, F. Francisco, P.J.S. Gil, J. Páramos, Thermal analysis of the Pioneer anomaly: a method to estimate radiative momentum transfer. Phys. Rev. D **78**(10), 103001 (2008). doi:10.1103/PhysRevD.78.103001
9. B. Bertotti, L. Iess, P. Tortora, A test of general relativity using radio links with the Cassini spacecraft. Nature **425**(6956), 374–376 (2003). doi:10.1038/nature01997
10. C.M. Will, The confrontation between general relativity and experiment. Living Rev. Relativ. **9** (2006). doi:10.12942/lrr-2006-3
11. L. Iess, G. Giampieri, J.D. Anderson, B. Bertotti, Doppler measurement of the solar gravitational deflection. Class. Quantum Gravity **16**(5), 1487–1502 (1999). doi:10.1088/0264-9381/16/5/303
12. O. Bertolami, F. Francisco, P.J.S. Gil, J. Páramos, Modeling the nongravitational acceleration during Cassini's gravitation experiments. Phys. Rev. D **90**(4), 042004 (2014). doi:10.1103/PhysRevD.90.042004
13. B. Christophe, L.J. Spilker, J. Anderson, N. André, S.W. Asmar, J. Aurnou, D. Banfield, A. Barucci, O. Bertolami, R. Bingham, P. Brown, B. Cecconi, J.M. Courty, H. Dittus, L.N. Fletcher, B. Foulon, F. Francisco, P.J.S. Gil, K.H. Glassmeier, W. Grundy, C. Hansen, J. Helbert, R. Helled, H. Hussmann, B. Lamine, C. Lämmerzahl, L. Lamy, R. Lehoucq, B. Lenoir, A. Levy, G. Orton, J. Páramos, J. Poncy, F. Postberg, S.V. Progrebenko, K.R. Reh, S. Reynaud, C. Robert, E. Samain, J. Saur, K.M. Sayanagi, N. Schmitz, H. Selig, F. Sohl, T.R. Spilker, R. Srama, K. Stephan, P. Touboul, P. Wolf, OSS (Outer Solar System): a fundamental and planetary physics mission to Neptune, Triton and the Kuiper Belt. Exp. Astron. **34**(2), 203–242 (2012). doi:10.1007/s10686-012-9309-y
14. J. Anderson, J. Campbell, J. Ekelund, J. Ellis, J. Jordan, Anomalous orbital-energy changes observed during spacecraft flybys of earth. Phys. Rev. Lett. **100**(9), 091102 (2008). doi:10.1103/PhysRevLett.100.091102
15. S.G. Turyshev, V. Toth, The puzzle of the flyby anomaly. Space Sci. Rev. **168**(1–4), 169–174 (2009). doi:10.1007/s11214-009-9571-0
16. O. Bertolami, F. Francisco, P.J.S. Gil, J. Páramos, Testing the flyby anomaly with the GNSS constellation. Int. J. Mod. Phys. D **21**, 1250035 (2012). doi:10.1142/S0218271812500356

Chapter 2
The Pioneer Anomaly and Thermal Effects in Spacecraft

2.1 General Background

The twin Pioneer 10 and Pioneer 11 deep-space probes were launched, respectively, on March 3, 1972 and April 6, 1973 as part of NASA's Pioneer program of planetary exploration. They were the first human objects to venture beyond the asteroid belt and provided invaluable scientific data from their visits to Jupiter and Saturn, including some of the first close up photographs of these planets, like those shown in Fig. 2.1 [1, 2].

The general configuration of these deep-space probes can be seen in Fig. 2.2 [3]. There is a main equipment compartment located directly behind the high-gain antenna, and two Radioisotope Thermal Generators (RTGs) mounted on two structures extending from the main compartment.

The scientific objectives of Pioneer 10 and 11 included the study of the interplanetary and planetary magnetic fields, solar wind parameters, cosmic rays, transition region of the heliosphere, neutral hydrogen abundance, and the atmosphere of Jupiter, Saturn and some of their satellites [4].

Pioneer 10 would become the first man-made object, by some definitions, to leave the solar system. The last time Pioneer 10 made contact with Earth was on January 23, 2003, from a distance of more than 80 AU. Communications with Pioneer 11 had already ceased after September 30, 1995. The two Pioneer probes currently follow hyperbolic trajectories away from the Solar System in approximately opposite directions [4], as shown in Fig. 2.3.

The success of these missions would pave the way for the heavier and more sophisticated Voyager 1 and 2 missions a few years later, that would repeat the visits to Jupiter and Saturn and be the first to make Uranus and Neptune flybys.

Through the later stages of the Pioneer 10 and 11 missions, the analysis of the radiometric data began to reveal the presence of an anomalous acceleration in the approximate direction of the Sun. The existence of this acceleration, that became known as the *Pioneer anomaly*, was first reported in 1998 in a paper by Anderson

F. Francisco, *Trajectory Anomalies in Interplanetary Spacecraft*, Springer Theses,
DOI 10.1007/978-3-319-18980-2_2

Fig. 2.1 Closeup pictures of Jupiter and Saturn taken, respectively, by Pioneer 10 and Pioneer 11 [1, 2]

Fig. 2.2 Artist's impression of one of the Pioneers heading into deep space [3]. The main components are clearly visible: the two RTGs and the main equipment compartment behind the main parabolic antenna

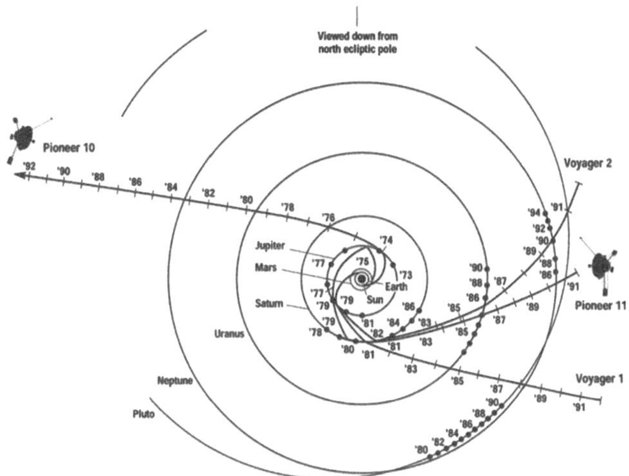

Fig. 2.3 Ecliptic pole view of Pioneer 10 and 11 and Voyager 1 and 2 trajectories. Taken from Ref. [5]

et al. [6]. The initial results pointed towards a constant value $a_{\text{Pioneer}} = (8.74 \pm 1.33) \times 10^{-10}\ \text{m/s}^2$.

The possibility of thermal effects being responsible for this anomaly was raised early in the debate about its cause in two comments to the initial Anderson et al. paper. In his comment, Murphy drew attention to the fact that waste heat from the electronic equipment was radiated from the spacecraft through a set of louvered radiators located on the side that faced away from the Sun and estimated that this would generate an acceleration between 3.2×10^{-10} and $8.5 \times 10^{-10}\ \text{m/s}^2$ [7]. Katz, on the other hand, mentioned that thermal radiation from the RTGs and scattered from the back of the antenna would be preferably directed away from the Sun, leading to a sunward acceleration [8]. Both argued, albeit on a qualitative basis, that thermal effects could provide significant clues for the unexplained acceleration.

In the more detailed study of the Pioneer anomaly that was presented by Anderson et al. in 2002, the thermal hypothesis was precluded. The authors asserted that RTG and electrical power would have decreased significantly in the observed period due to the fact that the primary power source was Plutonium-238, an isotope with a half-life of 87.7 years. This was not compatible with an acceleration that appeared from the Doppler data to be constant, as shown in Fig. 2.4. Furthermore, their initial assessment of systematic error sources indicated that any acceleration arising from heat dissipation would be too small to account for the Pioneer anomaly [5] (see also Ref. [9]).

Soon afterwards, a paper by Scheffer came out with a qualitative estimate of the thrust available from the different heat sources onboard the Pioneer spacecraft [10]. A breakdown of different effects was presented, each with the available power and an efficiency factor that represents the portion of this power that is converted into

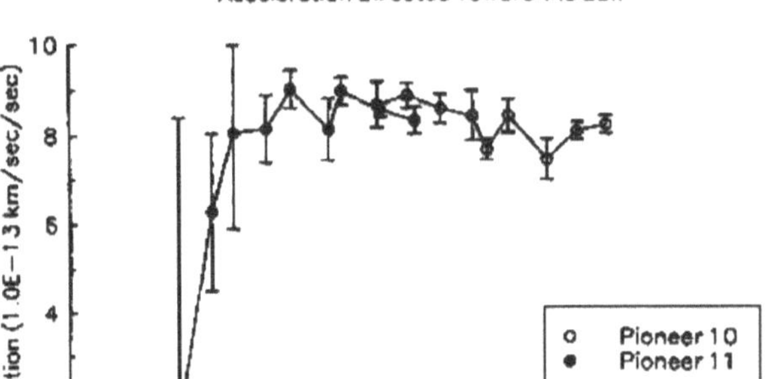

Fig. 2.4 Plot of the unmodelled accelerations of Pioneer 10 and 11, taken from Ref. [5]. Earlier data points do not represent true measurements, since they are masked by a large solar radiation force

Table 2.1 Scheffer's estimate of available thrust from different sources in terms of equivalent directed power, as shown in Ref. [10]

Source of effect	Total power (W)	Efficiency	Thrust (W)	Decay (%/year)
Radiation from RHUs	8	0.5	4	0.78
Antenna shadow	25	0.3	7.5	0.68
Antenna radiate	25	0.6	15	0.68
RTG asymmetry	2000	0.009	18	0.68
Feed pattern	0.8	0.7	0.6	0
Radio beam	7.2	−1	−7.2	0
Radiation, main bus	59	0.54	32	*cf.* Ref. [10]
Radiation, instruments	1	0.1	0.1	*cf.* Ref. [10]
Total			70	

thrust, as shown in Table 2.1, taken from Ref. [10]. The author estimated that 58 W of power directed away from the sun would be enough to account for the anomalous acceleration, while estimating a total available thrust of 70 W, thus concluding that it was likely that the whole effect could be explained without the need for new physics.

As this debate went on, numerous attempts were made to explain the Pioneer anomaly through new physics. A large array of proposals were made, ranging from scalar fields in the context of braneworld models [11]; theories of gravity with scalar, tensor and vector components [12], post-Einsteinian metric extensions of General

Relativity (GR) [13] to theories of gravity with non-minimal coupling between matter and curvature [14]. The first two of these are summarised in Sect. 2.2 as illustrative examples. A more complete review of the different proposals can be found in Ref. [15], including unsuccessful attempts to explain the anomaly resorting to more conventional physics, like the gravitational pull of the Kuiper Belt [16].

The crucial issue of the anomaly's time signature remained controversial for quite some time. The first independent confirmation of the existence of the anomaly was presented in an unpublished paper by Markwardt, where his own analysis tools were used [17]. The best fit obtained from the Doppler data was a constant acceleration of $(8.60 \pm 1.34) \times 10^{-10}$ m/s^2. However, the author pointed out that this result was also statistically compatible with the time signature of the radioactive decay of the Plutonium-238 powering the RTGs and, as a consequence, a relation between the anomalous acceleration and heat dissipation could not be excluded.

At least two other independent Doppler data processing efforts were performed, one by Levy et al. [18] and another by Toth [19], further reinforcing confidence on the existence of a Pioneer anomaly. Besides both confirming the existence of a constant component with values similar to the previously reported, the former focuses its analysis on short time-scale periodic accelerations, while the latter includes the estimate of a "jerk" term (time derivative of the acceleration), finding it to be consistent with the expected temporal variation of a recoil force due to heat generated on board and emitted anisotropically [19].

The issue was finally settled in a paper co-authored by Turyshev, a member of the team at the Jet Propulsion Laboratory (JPL) that originally put forward the existence of the anomaly, with the collaboration of Markwardt, Ellis and Toth. The results supported a time-decaying acceleration compatible with a thermal origin to the anomaly [20]. The tracking data was fitted to constant, linear and exponential models with the latter having the smaller residuals, as summarised in Table 2.2 and Fig. 2.5 [20].

At this point, it had long became clear that a reliable quantitative description of the effects of Pioneer's thermal emissions was essential. In response to this need, three independent studies were carried out in the last few years by scientific teams working in Portugal, Germany and the United States.

Table 2.2 Pioneer 10 and 11 acceleration fits to constant, linear, and exponential models, as reported in Ref. [20]

Spacecraft	Model	σ_p (mHz)	$a_P(t_0)$ (10^{-10} m/s^2)	$\dot{a}_P$ (m/s^2/year)	β^{-1} (year)
Pioneer 10	Constant	4.98	8.17	–	–
	Linear	4.60	11.06	−0.17	–
	Exponential	4.58	12.22	–	28.8
Pioneer 11	Constant	3.67	9.15	–	–
	Linear	2.09	11.65	−0.18	–
	Exponential	2.06	13.79	–	24.6

σ_p are the root mean square residuals, $a_P(t_0)$ is the acceleration on January 1, 1972, $\dot{a}_P$ is the slope in the linear model and β^{-1} is the half-life in the exponential model

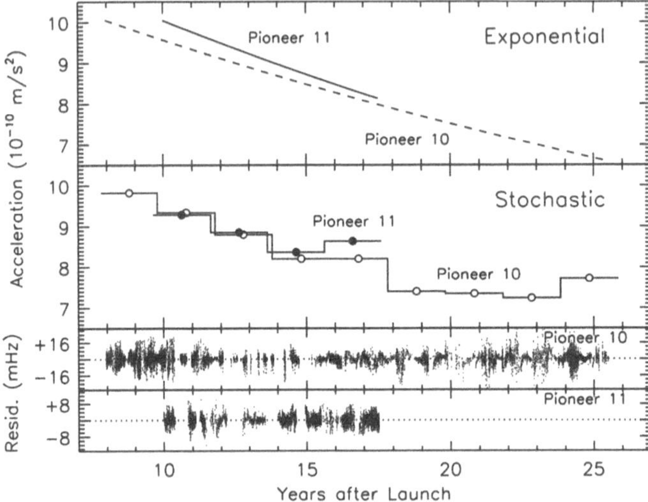

Fig. 2.5 *Top panel* Estimates of the anomalous acceleration of Pioneer 10 (*dashed line*) and Pioneer 11 (*solid line*) using an exponential model. *Second panel* Stochastic acceleration estimates for Pioneer 10 (*open circles*) and Pioneer 11 (*filled circles*), shown as step functions. *Bottom two panels* Doppler residuals of the stochastic acceleration model. Note the difference in vertical scale for Pioneer 10 versus Pioneer 11, show that Pioneer 10 data is much noisier. Taken from Ref. [20]

This chapter of the thesis sums up the work made by the Portuguese team in tackling this problem. The first results appeared in 2008, indicating that up to 67 % of the acceleration was explainable through thermal effects [21]. The results were obtained using a novel method based on a distribution of point-like radiation sources, that was thoroughly tested to ensure its reliability and conformity with the objectives that were set out [21, 22]. In order to provide a full description of the physics involved, the method was subsequently extended to include the modelling of reflections [23], as explored in detail in Sect. 2.3.

As we shall see, the approach chosen by our team allowed for the first complete and reliable determination of the thermal acceleration aboard the Pioneer and to conclude that they were, in fact, responsible for the Pioneer anomaly [23].

The results were subsequently confirmed by the German team working at the Center for Applied Space Technology and Microgravity (ZARM) in Bremen [24] and, finally, by the American team at the JPL [25], both using more conventional finite-element models.

2.2 New Physics Proposals

Out of the many attempts to explain the Pioneer anomaly through new physics, we briefly explore two illustrative examples of the kind of proposals that were put forward.

2.2.1 Scalar Field

One possible way to look at the problem of the Pioneer anomaly is in the context of a braneworld scenario, as done by Bertolami and Páramos [11]. In braneworld theories, our Universe is assumed to be a 4-dimensional membrane embedded in a higher dimensional bulk space.

The authors first attempt to use a Randall-Sundrum braneworld model and explain the Pioneer anomaly as an influence of the radion field, a scalar perturbation of the metric related to relative motion of the two branes. The authors conclude that this approach is unsuitable to provide the desired explanation.

A more promising possibility can arise from the presence of a scalar field ϕ with a potential $V(\phi) \propto -\phi^{-\alpha(r)}$, with $\alpha > 0$. This field is similar to the form a supergravity inspired quintessence field assumes in braneworld theories, but with the sign reversed.

The effect appears in the metric $g_{\mu\nu}$ as a small perturbation $h_{\mu\nu}$ to the Minkowsky metric $\eta_{\mu\nu}$, so that

$$g_{\mu\nu} = \eta_{\mu\nu} + h_{\mu\nu}. \tag{2.1}$$

The Lagrangian density $\mathcal{L}_\phi$ of the scalar field takes the form

$$\mathcal{L}_\phi = \frac{1}{2}\eta_{\mu\nu}\partial_\mu\phi\partial^\nu\phi - V(\phi). \tag{2.2}$$

Since this is a spherically symmetric problem, the formulation can be developed in spherical coordinates. Hence, Eq. (2.2) takes the form

$$\mathcal{L}_\phi = \frac{1}{2}\eta_{rr}(\phi')^2 - A^2\phi^{-\alpha}, \tag{2.3}$$

where A is a constant. The equation of motion of the scalar field is

$$\Box^2\phi + \frac{dV(\phi)}{d\phi} = 0, \tag{2.4}$$

where $\Box = \partial_\mu\partial^\mu$ is the d'Alembertian operator. This equation has as a solution in spherical coordinates

$$\phi(r) = \left((2+\alpha)\sqrt{\frac{\alpha}{8+\alpha}}Ar\right)^{\frac{2}{2+\alpha}} \equiv \beta^{-1}r^{\frac{2}{2+\alpha}}. \tag{2.5}$$

This means that the potential takes the form

$$V(\phi(r)) = -A^2\beta^\alpha r^{-\frac{2}{2+\alpha}}, \tag{2.6}$$

while the gradient term becomes

$$\frac{1}{2}(\phi'(r))^2 = A\left(\frac{\alpha}{4+\alpha}\right)\beta^\alpha r^{-\frac{2\alpha}{2+\alpha}}. \tag{2.7}$$

The Lagrangian density in the Newtonian limit is given by

$$\mathcal{L}_\phi = -\frac{4}{4+\alpha}V(\phi). \tag{2.8}$$

The energy-momentum tensor for the scalar field is given by the expression

$$T_{\mu\nu} = \partial_\mu\phi\partial_\nu\phi - \eta_{\mu\nu}\mathcal{L}_\phi. \tag{2.9}$$

This is then introduced in the linearised form of Einstein's equation

$$\frac{1}{2}\nabla^2 h_{\mu\nu} = 8\pi G\left(T_{\mu\nu} - \frac{1}{2}\eta_{\mu\nu}T\right), \tag{2.10}$$

where G is the gravitational constant. From the solution of this equation, one can obtain the radial acceleration caused by the scalar field

$$a_r = -\frac{C}{r^2} + (2+\alpha)A^2 8\pi G\beta^\alpha r^{-\frac{2\alpha}{2+\alpha}}\left(\frac{C}{2} - \frac{r}{6+\alpha}\right), \tag{2.11}$$

where C is a constant.

For $\alpha = 2$ one gets an expression for the acceleration that is compatible with the main constant observational signature of the Pioneer Anomaly:

$$a_r = -\frac{C}{r^2} + \sqrt{\frac{3}{2}}A^2 2\pi G + \sqrt{\frac{3}{2}}\frac{AC8\pi G}{r}. \tag{2.12}$$

The first term represents the Newtonian contribution and the term proportional to r^{-1} is much smaller than the constant term for $4C/r \ll 1$, that is for $r \gg 6$ km, and is also much smaller than the Newtonian acceleration for $r \ll 2.9 \times 10^{22}$ km ≈ 100 Mpc, clearly covering the desired range. The constant term can, therefore, be identified with the anomalous acceleration by setting the constant A appropriately, for instance, if $a_{\text{Pioneer}} = 8.5 \times 10^{-10}$ m/s^2 then $A = 4.7 \times 10^{42}$ m^{-3} [11].

2.2.2 Scalar-Tensor-Vector Gravity

Another proposal was put forward by Brownstein and Moffat using Scalar-Tensor-Vector Gravity (STVG) theory to obtain an effect that fits the available data. The theory is outlined in Ref. [12] and postulates the existence of a spin-1 vector field ϕ. Furthermore, in this theory the gravitational constant G, the vector field coupling strength ω and the vector field mass $\mu = 1/\lambda$ are all treated as scalar fields with their own dynamics. The action for STVG takes the form

$$S = \int dx^4 \sqrt{-g}(\mathcal{L}_{\text{Grav}} + \mathcal{L}_\phi + \mathcal{L}_S), \tag{2.13}$$

where g is the determinant of the metric. This action includes the Lagrangian densities for the vector field

$$\mathcal{L}_\phi = \omega \left(\frac{1}{4}(\partial^\mu \phi^\nu - \partial^\nu \phi^\mu)(\partial_\mu \phi_\nu - \partial_\nu \phi_\mu) + V(\phi) \right), \tag{2.14}$$

for the scalar quantities G, ω and μ

$$\begin{aligned} \mathcal{L}_S = & \frac{1}{G^3}\left(\frac{1}{2}g^{\mu\nu}\nabla_\mu G \nabla_\nu G + V(G)\right) \\ & + \frac{1}{G}\left(\frac{1}{2}g^{\mu\nu}\nabla_\mu \omega \nabla_\nu \omega + V(\omega)\right) \\ & + \frac{1}{\mu^2 G}\left(\frac{1}{2}g^{\mu\nu}\nabla_\mu \mu \nabla_\nu \mu + V(\mu)\right) \end{aligned} \tag{2.15}$$

and for gravitation

$$\mathcal{L}_{\text{grav}} = \frac{1}{16\pi G}(R + 2\Lambda), \tag{2.16}$$

where R is the scalar curvature and Λ is the cosmological constant.

From the development of the field equations, the equations of motion for a static spherically symmetric field about a central mass M can be obtained. The line element is written in spherical coordinates as

$$ds^2 = \gamma(r)dt^2 - \alpha(r)dr^2 - r^2(d\theta^2 + \sin^2\theta d\varphi^2). \tag{2.17}$$

An exact solution for the spherically symmetric static field equations can be obtained if the potential $V(\phi)$ and Λ are small enough to be neglected, yielding

$$\gamma(r) = 1 - \frac{2GM}{r} + \frac{Q^2}{r^2}, \quad \alpha(r) = 1 - \frac{2GM}{r} + \frac{Q^2}{r^2}, \tag{2.18}$$

where the charge ϵ of the spin-1 vector particle is taken into account in the quantity

$$Q = 4\pi G\omega\epsilon^2. \tag{2.19}$$

That can be compared to the usual Schwarzschild solution

$$\gamma_{\text{Schwarz}}(r) = 1 - \frac{2GM}{r}, \quad \alpha_{\text{Schwarz}}(r) = \frac{1}{1 - \frac{2GM}{r}}. \tag{2.20}$$

We can easily see that, as expected, for large values of r the STVG solution degenerates in the Schwarzschild solution. With some more manipulation, described in detail in Ref. [12], we finally obtain the equation of motion of a particle around a mass M

$$\frac{d^2r}{dt^2} - \frac{J_N^2}{r^3} + \frac{GM}{r^2} = K\frac{e^{-\mu r}}{r^2}(1 + \mu r). \tag{2.21}$$

where J_N is the Newtonian orbital angular momentum and K is a positive quantity. Following the formulation developed in Ref. [12], the radial acceleration can be written as

$$a(r) = -\frac{G_\infty M}{r^2} + K(r)\frac{e^{-r/\lambda(r)}}{r^2}\left(1 - \frac{r}{\lambda(r)}\right). \tag{2.22}$$

The value for the gravitational constant appears renormalized as

$$G_\infty = G_0(1 + \alpha_\infty), \tag{2.23}$$

where G_0 here denotes the Newtonian gravitational constant. The value for K is chosen as

$$K(r) = G_0 M \alpha(r). \tag{2.24}$$

Using Eq. (2.24), we can finally write the variation of G with distance to the central mass

$$G(r) = G_0\left[1 + \alpha(r)\left[1 - e^{-r/\lambda(r)}\left(1 - \frac{r}{\lambda(r)}\right)\right]\right] \tag{2.25}$$

and the acceleration

$$a(r) = -\frac{G(r)M}{r^2}. \tag{2.26}$$

The authors then postulate that the Pioneer Anomaly is caused by the difference between the Newtonian gravitational constant G_0 and the new dynamic value $G(r)$. The anomalous Pioneer acceleration would thus be given by

$$a_{\text{Pio}} = -\frac{\Delta G(r)\, M_\odot}{r^2}, \tag{2.27}$$

where

$$\Delta G(r) = G(r) - G_0 = G_0\left[\alpha(r)\left[1 - e^{-r/\lambda(r)}\left(1 - \frac{r}{\lambda(r)}\right)\right]\right]. \tag{2.28}$$

The proposed parametric representations for $\alpha(r)$ and $\lambda(r)$ are:

$$\alpha(r) = \alpha_\infty (1 - e^{\frac{r}{\bar{r}}})^{\frac{b}{2}}, \tag{2.29}$$

$$\lambda(r) = \frac{\lambda_\infty}{(1 - e^{\frac{r}{\bar{r}}})^b}. \tag{2.30}$$

Here, $\bar{r}$ is a non-running distance scale parameter and b is a constant.

Using a least-squares routine, the authors obtain values for the constant parameters that yield the best fit to the acceleration residuals:

$$\alpha_\infty = (1.00 \pm 0.02) \times 10^{-3}, \tag{2.31}$$

$$\lambda_\infty = 47 \pm 1\text{AU}, \tag{2.32}$$

$$\bar{r} = 4.6 \pm 0.2\text{AU}, \tag{2.33}$$

$$b = 4.0. \tag{2.34}$$

The graph in Fig. 2.6 plots the obtained prediction for the anomalous acceleration compared with the data from both Pioneer probes.

Finally, the authors argue that the STVG theory can explain the anomalous acceleration and still be consistent with the equivalence principle, lunar laser ranging and satellite data for the inner solar system as well as the outer solar system planets [12].

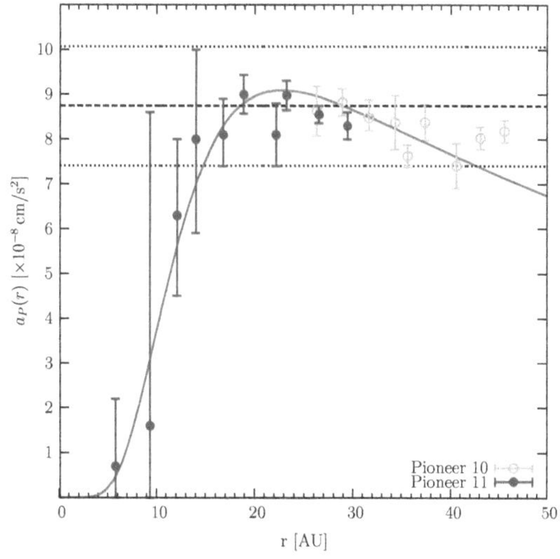

Fig. 2.6 Best fit to the Pioneer anomalous acceleration data plotted against the position, r in AU, on a linear scale out to $r = 50$ AU, as presented in Ref. [12]. Compare with Fig. 2.4

2.3 Point-Like Source Method

2.3.1 Motivation

Even during the most heated part of the debate, every assessment of the anomalous acceleration left open the effect due to thermal radiation. The implication was that thermal effects became a controversial issue, which meant that a reliable estimate of these effects was key to the understanding of the problem.

The more conventional approach would be to construct a detailed finite-element model of the spacecraft and, based on individual instrument power consumptions, produce a map of external surface temperatures from which thermal radiation would be derived. This was the path followed by the teams working out the issue at ZARM [24] and the JPL [25].

The main disadvantage of these methods in this context is their reliance on a very accurate knowledge of all the engineering details involved, including structural configuration and dimensions, thermal and optical properties of the materials used and individual instrument power consumptions. Much of this information was not easily available, meaning that any effort to model Pioneer 10 and 11 through these methods would necessarily involve a significant amount of guessing. Moreover, the finite-element models are too slow to allow for the testing of a wide enough range of values for the poorly known parameters. Up until recently, there was also no unequivocal characterisation of the acceleration, implying that the efficiency of such an effort would be limited.

This problem required an approach that, not only kept the physical formulation transparent, but also combined a high degree of flexibility and computational speed.

The choice made by our team was to develop a new approach that would be able to tackle this issue with all its challenges. We called this the *point-like source method*, due to one of its key features, the modelling of the spacecraft's thermal emissions through a distribution of a small number of carefully placed point-like radiation sources.

In order to ensure that simplicity was not achieved at the expense of accuracy, a battery of test cases was performed to evaluate the robustness of the results. The outcome of these test cases, discussed in Sect. 2.3.3, validates the approach by showing that, for the typical configurations of the surfaces in the spacecraft, the result converges rapidly and with a relatively small number of point-like sources [22].

One key feature of the point-like source method is that it easily allows for the variation of parameters involving a large degree of uncertainty. This is related to the geometrical and material properties of the various spacecraft elements, which, in most cases, do not have well-known values, even at launch, and have endured extended periods of degradation in space. We have used a Monte Carlo simulation, by assigning a statistical distribution to the values of each parameter, based on the available information, and generating a large number of random values to obtain a probability distribution for the final result [23].

The fact that this method produced results that are generally in agreement with the ones obtained through subsequent, more detailed finite-element models [24, 25], is a further indication of its reliability and robustness. This was further demonstrated by its successful application to the study of the Cassini spacecraft, discussed in Chap. 3 of this thesis [26].

2.3.2 Radiative Momentum Transfer

The description of the emitted radiation, as well as its reflection, is always made in terms of the the Poynting vector, which represents the energy flux. For instance, the time-averaged Poynting vector field for a Lambertian source located at $\mathbf{x}_0$ is given by

$$\mathbf{S}_{\mathrm{Lamb}}(\mathbf{x}) = \frac{W}{\pi||\mathbf{x}-\mathbf{x}_0||^2}\left(\mathbf{n}\cdot\frac{\mathbf{x}-\mathbf{x}_0}{||\mathbf{x}-\mathbf{x}_0||}\right)\frac{\mathbf{x}-\mathbf{x}_0}{||\mathbf{x}-\mathbf{x}_0||}, \tag{2.35}$$

for $\mathbf{n}\cdot(\mathbf{x}-\mathbf{x}_0)\geq 0$, where W is the emissive power and $\mathbf{n}$ is the emitting surface normal.

Despite its point-like nature, the method can be easily extended to include other radiation source geometries, as long as they have a straightforward mathematical description and preserve energy conservation. An example that turned out to be especially useful in the thermal modelling of Cassini [26] is the cylindrical source, where the emitter is a line segment instead of a point and the Poynting vector field has cylindrical symmetry. For example, one can write the time averaged Poynting vector field of a cylindrical source parallel to the x-axis with coordinates (y_0, z_0) in the yz-plane as

$$\mathbf{S}_{\mathrm{cyl}}(\mathbf{x}) = \frac{W(0, y-y_0, z-z_0)}{2\pi l((y-y_0)^2+(z-z_0)^2)}, \tag{2.36}$$

where W is the emissive power, l is the length of the source and $\mathbf{x}=(x, y, z)$.

From the Poynting vector field, the amount of power illuminating a given surface, W_{illum}, can be obtained by computing the Poynting vector flux through that surface, given by the integral

$$W_{\mathrm{illum}} = \int_S \mathbf{S}\cdot\mathbf{n}_{\mathrm{illum}}\, dA, \tag{2.37}$$

where $\mathbf{n}_{\mathrm{illum}}$ is the illuminated surface normal and $\mathbf{S}$ is the Poynting vector field resulting from the sum of all relevant radiation sources.

Since electromagnetic radiation carries momentum, its emission or absorption will translate into a force being exerted on illuminated surfaces. This is usually expressed as a *radiation pressure*, P_{rad}, given, for an opaque illuminated surface by the power flux divided by the speed of light

$$P_{\text{rad}} = \frac{\mathbf{S} \cdot \mathbf{n}_{\text{illum}}}{c}. \tag{2.38}$$

In the case of a radiation source, the radiation pressure has its sign reversed relative to the direction of emission. If there is transmission (i.e., the surface is not opaque) the pressure is multiplied by the absorption coefficient. As to reflection, we shall see in Sect. 2.3.4 that it can be treated as an absorption followed by a partial re-emission of the radiation.

Integrating the radiation pressure on a surface allows us to obtain the force and, dividing by the mass of the spacecraft, m_{sc}, the acceleration due to thermal radiation

$$\mathbf{a}_{\text{th}} = \frac{1}{m_{\text{sc}}} \int_S \frac{\mathbf{S} \cdot \mathbf{n}_{\text{illum}}}{c} \frac{\mathbf{S}}{||\mathbf{S}||} dA. \tag{2.39}$$

To determine the force exerted by the radiation on the emitting surface, the integral should be taken over a closed surface encompassing the latter in order to determine the total momentum carried by the radiation. Equivalently, the force exerted by the radiation on an illuminated surface requires an integration surface that encompasses it. Furthermore, considering a set of emitting and illuminated surfaces implies the proper accounting of the effect of the shadows cast by the various surfaces, which are then subtracted from the estimated force on the emitting surface. One may then read the thermally induced acceleration directly.

It should be highlighted at this point that the formulation here presented relies solely on the power balance of the spacecraft's external surfaces and there is no intermediate step where the temperatures are obtained. This is important because it shortcuts the uncertainty related to the fragmented and incomplete data on surface emissivities and the limited onboard temperature telemetry data. As discussed further on, in Sect. 2.4.5, the data on the power supply is much more reliable.

2.3.3 *Test Cases*

The innovative character of this method makes it wise to conduct a series of test cases to confirm if it delivers the correct results. The purpose is to assess the quality of the results and to gain sensitivity to the errors involved in the kind of approximations that are performed. The main issue is to ascertain if the radiation emitted from an extended surface can be adequately represented by a small number of point-like sources instead of a very fine mesh of radiating elements.

The test case setup includes a 1 m^2 emitting surface and a second absorbing surface of similar size set at various distances and angles. These characteristic sizes and distances were chosen to be of the same order of magnitude as those involved in the construction of the kind of spacecraft under analysis. The radiation emissions are modelled with an increasingly finer mesh of point-like Lambertian sources and the results for the energy flux and force are then compared and analysed for convergence.

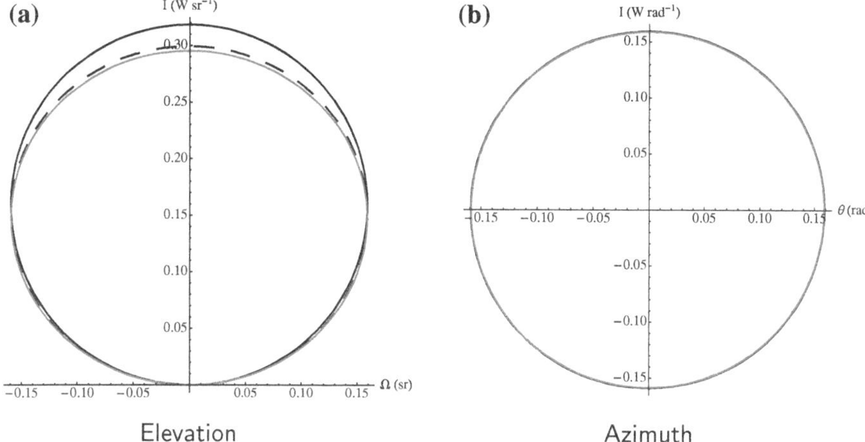

Fig. 2.7 Polar plot of the energy flux variation with elevation and azimuth of the radiation emitted by a surface on the xy-plane, when considering 1, 4 and 16 Lambertian sources (*full*, *dashed* and *grey curves*, respectively), maintaining the total emitted power constant at 1 W. The curves for 64 or 144 sources overlap the one for 16 sources). The intensity at higher elevations (close to vertical) diminishes with the number of sources, compensating the slight increase at the lower angles

For a single radiation emitting surface without any other illuminated surfaces, the force is normal to it and only depends on the total emitted power. Integrating Eq. (2.39) along a closed surface encompassing the emitting surface in the xy-plane, we obtain a force pointing in the z-axis, of magnitude $(2/3)W/c$, where W is the emitted power and c is the speed pf light.

Computation of the shadow and radiation pressure on a second surface yields results that are not independent from the source distribution. In order to acquire some sensibility on that dependence, we plot the variation of the radiation intensity with the elevation and the azimuth for 1, 4, 16, 64 and 144 source meshes, as depicted in Fig. 2.7. These plots are obtained by integrating the energy flux along the azimuth or elevation.

A visual inspection of the results indicates that, even for the simpler 1 source mesh, the maximum deviation occurs at the higher angles of elevation and is less than 10 %, when compared to the 144 source mesh. Except in cases where surfaces directly face each other, deviations should be considerably smaller. One also verifies that convergence is achieved very rapidly, since the intensity plots for 16, 64 and 144 source meshes are superimposed, having minimal differences between them.

In order to confirm this intuitive perception, the force acting on a second 1 m^2 surface is computed for several different positions. A total of nine configurations were tested, with different positions and tilt angles, as listed in Table 2.3. Two of the configurations are also illustrated in Fig. 2.8 as examples. The results are then obtained for 1, 4, 16, 64 and 144 source meshes. The configurations were chosen to be representative of the typical dimensions and angles involved in the geometric

Table 2.3 Positions considered for the illuminated surface in test cases

Test case	Surface centre position (m)	Surface tilt angle (°)
1	(2, 0, 0.5)	90
2	(2, 0, 1.5)	0
3	(2, 0, 1.5)	30
4	(2, 0, 1.5)	60
5	(2, 0, 1.5)	90
6	(1, 0, 2)	0
7	(1, 0, 2)	30
8	(1, 0, 2)	60
9	(1, 0, 2)	90

The first (emitting) surface is in the xy plane centred at the origin. Considered distances between both surfaces are typical for the Pioneer spacecraft

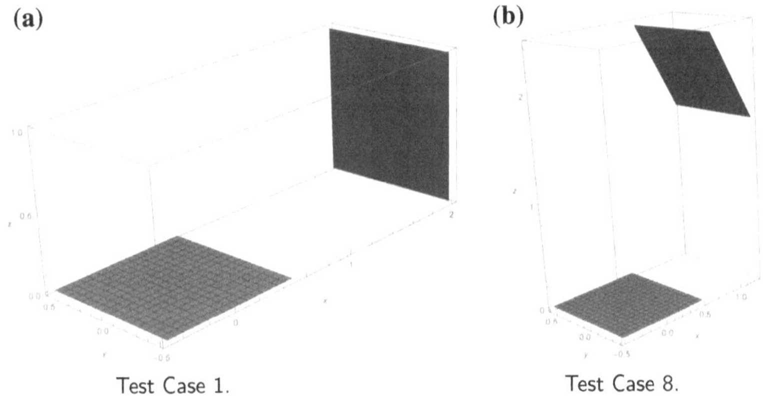

Fig. 2.8 Geometry of test cases 1 and 8 (*cf.* Table 2.3). Thermal emission from a surface is simulated by a different number of Lambertian sources evenly distributed on the surface, maintaining the total emitted power constant, and the effect on a second surface is observed. This is the test case where the highest variation with the number os sources considered were obtained

configurations of the kind of spacecraft in question. The full results for the energy flux and force components for each test case are presented in Appendix A.

Of all the analysed cases, the highest deviation occurs for Test Case 8, confirming previous expectations, since the second surface is set at high elevation from the emitting surface, as depicted in Fig. 2.8b. The results in Table 2.4 show a difference of approximately 6 % between the force obtained with one source and the results for the finer meshes (16, 64 and 144 sources). Nevertheless, the latter are all within 0.5 % of each other, and the intermediate 4 source mesh has a deviation of under 1.5 % to the 144 source mesh.

Table 2.4 Results for test case 8 considering a total emission of 1 kW

Sources	Energy Flux (W)	Δ (%)	Force components (x, y, z) $(10^{-7}$ N)	Force intensity $(10^{-7}$ N)	Δ (%)
1	45.53		(2.016, 0, 2.083)	2.899	
4	43.85	3.7	(1.918, 0, 2.003)	2.773	4.3
16	43.45	0.93	(1.895, 0, 1.984)	2.744	1.1
64	43.35	0.23	(1.890, 0, 1.979)	2.736	0.27
144	43.33	0.043	(1.889, 0, 1.978)	2.735	0.050

As the number of sources to represent the thermal emission of a surface change, the resultant force components on the secondary surface remain almost the same

Table 2.5 Same as Table 2.4, for test case 1

Sources	Energy Flux (W)	Δ (%)	Force components (x, y, z) $(10^{-7}$ N)	Force intensity $(10^{-7}$ N)	Δ (%)
1	15.34		(0.9300, 0, 0.1514)	1.004	
4	15.92	3.8	(1.028, 0, 0.1638)	1.041	3.6
16	16.09	1.0	(1.038, 0, 0.1675)	1.051	0.98
64	16.13	0.26	(1.040, 0, 0.1684)	1.054	0.25
144	16.14	0.049	(1.041, 0, 0.1686)	1.054	0.047

For the typical angles of the most common space probe configurations, one may take as figure of merit Test Case 1, depicted in Fig. 2.8a, with the results shown in Table 2.5. The analysis of these results shows that, for 16, 64 and 144 sources, the variation in the energy flux and force is, again less than 0.5 %. In addition to that, the difference to the finer meshes is less than 5 % for 1 source and less than 1.5 % for a 4 source mesh.

An inspection of the remaining cases listed in Appendix A yields similar conclusions. For all test cases examined, the convergence of the results occurs at a similar pace and yields similarly small deviations. These results provide a fairly good illustration of the power of the proposed method and how well we can estimate the radiation effects on the Pioneer probes and other problems with similar requirements. The deviations are always well below 10 %, even with the roughest simplifications allowed by the method. We may then conclude that, for the scales and geometry involved in a thermal model of a space probe such as the Pioneer 10 and 11, the source distribution method is, not only consistent and convergent, but that it provides a very satisfactory account of the thermal radiation effects, considering all uncertainties involved.

After analysing the convergence of the method, we also wanted to assess the effect of ignoring minor surface features, such as the equipment attached to the external walls of the spacecraft and other geometric details. For that we considered

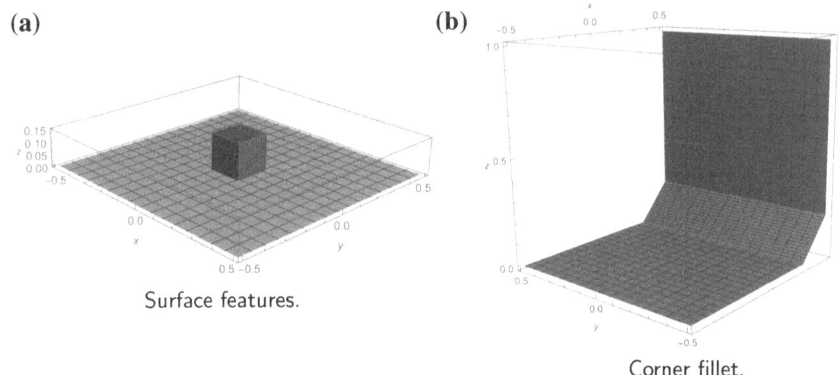

Fig. 2.9 Geometry for the surface features and corner fillet test cases. A cubical shape is placed on top of the flat surface and the force is compared for different sizes of this cube, while the total power is kept constant

Table 2.6 Results for the surface feature test case where the impact of ignoring a cubic shape placed on top of a 1 m^2 flat surface is analysed

Feature height (cm)	Force intensity (10^{-6} N)	Δ (%)
0	2.224	
1	2.223	0.040
5	2.202	1.0
10	2.139	3.9

The total power is kept constant at 1 kW and the temperature is assumed uniform in all surfaces. The deviations Δ with respect to the plane surface without any features are small enough to allow this simplification

two additional test cases. The particular situations analysed were a cubical piece on top of a 1 m^2 flat surface, as shown in Fig. 2.9a, and two perpendicular surfaces with a fillet (a "cut corner"), as in Fig. 2.9b. In each case, the force resulting from the emissions of the surfaces was compared. The total power is kept constant and the temperature is assumed uniform in all surfaces.

The results presented in Tables 2.6 and 2.7 set boundaries on the kind of geometric simplifications that can be made without a significant impact in the final result and keeping in line with the accuracy targets set for this study.

These results indicate that, in the absence of large temperature gradients, no significant errors will arise from considering flat surfaces and not taking into account all the minute details of the spacecraft.

Table 2.7 Results for the corner fillet test case with a constant total power of 1 kW

Fillet dimension (cm)	Force components (x, y, z) $(10^{-7}$ N)	Force intensity $(10^{-7}$ N)	Δ (%)
0	$(1.112, -1.112, 0)$	1.573	
1	$(1.115, -1.115, 0)$	1.577	0.2
5	$(1.129, -1.129, 0)$	1.596	1.5
10	$(1.146, -1.146, 0)$	1.620	3.0
20	$(1.181, -1.181, 0)$	1.670	6.2

The deviation Δ of the force intensity with respect to the sharp corner are kept within reasonable values

2.3.4 Reflection Modelling: Phong Shading

The inclusion of reflections in the model is achieved through a method known as *Phong Shading*, a set of techniques and algorithms commonly used to render the illumination of surfaces in three-dimensional computer graphics [27].

This methodology includes a reflection model which accounts for diffusive and specular reflection, known as *Phong reflection model*, and an interpolation method for curved surfaces modelled as polygons, known as *Phong interpolation.*

The Phong reflection model is based on an empirical formula that gives the illumination value of a given point in a surface, I_{p}, as

$$I_{\mathrm{p}} = k_{\mathrm{a}} i_{\mathrm{a}} + \sum_{m \in \mathrm{lights}} \left[k_{\mathrm{d}} (\mathbf{l}_m \cdot \mathbf{n}) i_{\mathrm{d}} + k_{\mathrm{s}} (\mathbf{r}_m \cdot \mathbf{v})^{\alpha} i_{\mathrm{s}} \right], \tag{2.40}$$

where k_{a}, k_{d} and k_{s} are the ambient, diffusive and specular reflection constants, i_{a}, i_{d} and i_{s} are the respective light source intensities, $\mathbf{l}_m$ is the direction of the light source m, $\mathbf{n}$ is the surface normal, $\mathbf{r}_m$ is the direction of the reflected ray, $\mathbf{v}$ is the direction of the observer and α is a "shininess" constant (the higher it is, the more mirror-like is the surface).

In using this formulation to solve a physics problem, there are a few constrains that should be taken into account. The ambient light parameter k_{a} and i_{a}, while useful in computer graphics, are not relevant for this problem since they give the reflection behavior relative to a background radiation source. Also, the intensities i_{d} and i_{s} should be the same, since the diffusive and specular reflection are relative to the same radiation sources.

This method provides a simple and straightforward way to model the various components of reflection, as well as an accurate accounting of the thermal radiation exchanges between the surfaces on the spacecraft. In principle, there is no difference between the treatment of thermal infrared radiation and visible light, for which the method was originally devised, allowing for a natural wavelength dependence of the above material constants.

Bearing in mind the formulation used in Sect. 2.3.2, the Phong shading methodology must be adapted from a formulation based on intensities (energy per unit surface of the projected emitting surface) to one based on the energy-flux per unit surface (the Poynting vector). This led to two expressions for the diffusive and specular components of reflection in terms of the Poynting vector field. Thus, the diffusively reflected radiation Poynting vector field is given by

$$\mathbf{S}_{\mathrm{rd}}(\mathbf{x}, \mathbf{x}') = \frac{k_{\mathrm{d}}|\mathbf{S}(\mathbf{x}') \cdot \mathbf{n}|}{\pi ||\mathbf{x} - \mathbf{x}'||^2} \mathbf{n} \cdot (\mathbf{x} - \mathbf{x}') \frac{\mathbf{x} - \mathbf{x}'}{||\mathbf{x} - \mathbf{x}'||}, \tag{2.41}$$

while the specular component reads

$$\mathbf{S}_{\mathrm{rs}}(\mathbf{x}, \mathbf{x}') = \frac{k_{\mathrm{s}}|\mathbf{S}(\mathbf{x}') \cdot \mathbf{n}|}{\frac{2\pi}{1+\alpha} ||\mathbf{x} - \mathbf{x}'||^2} [\mathbf{r} \cdot (\mathbf{x} - \mathbf{x}')]^{\alpha} \frac{\mathbf{x} - \mathbf{x}'}{||\mathbf{x} - \mathbf{x}'||}, \tag{2.42}$$

where $\mathbf{x}'$ is a point on the reflecting surface. In both cases, the reflected radiation field depends on the incident radiation field $\mathbf{S}(\mathbf{x}')$ and on the reflection coefficients k_{d} and k_{s}, respectively.

Using Eqs. (2.41) and (2.42), one can now compute the reflected radiation field by adding up both components. From the emitted and reflected radiation vector fields, the irradiation of each surface is then computed and, from that, the contribution to the acceleration can be obtained through Eq. (2.39).

To summarise, the first step in the procedure is, once the radiation source distribution is put in place, to compute the emitted radiation field and the respective force exerted on the emitting surfaces. This is followed by the determination of which surfaces are illuminated and the computation of the force exerted on those surfaces by the radiation. At this stage, we get a figure for the thermal force without reflections. The reflection radiation field is then computed for each surface and subject to the same steps as the initially emitted radiation field, leading to a determination of thermal force with one reflection.

This method can, in principle, be iteratively extended to as many reflection steps as desired, considering the numerical integration algorithms and available computational power. If necessary, the computation time in each step can be reduced through a discretisation of the reflecting surface into point-like reflectors.

2.4 Pioneer Thermal Model

2.4.1 Geometric Model

The construction of a geometric model of the Pioneer space probes that follows the principles outlined in the previous section is the first task to perform in the study of Pioneer 10 and 11 thermal effects.

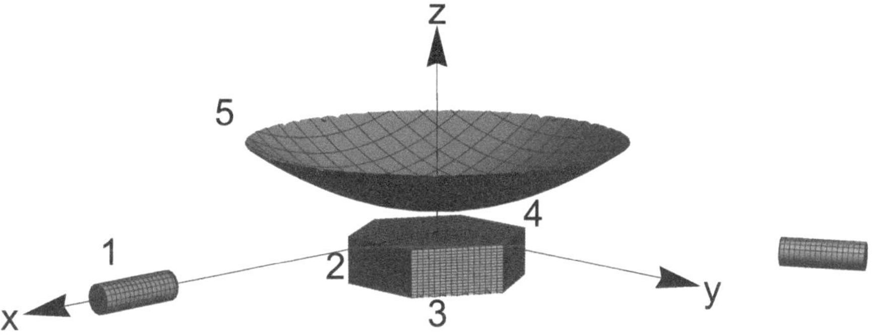

Fig. 2.10 Three-dimensional simplified geometric model of Pioneer 10 and 11 showing the configuration of the RTGs (*1*), the mai equipment compartment with its side (*2*), front (*3*) and back walls (*4*), and the parabolic high-gain antenna (*5*)

The geometric model used in this study retains the most important features of the Pioneer spacecraft, namely:

(i) the parabolic high-gain antenna,
(ii) the main equipment compartment behind the antenna,
(iii) two Radioisotope Thermal Generators (RTGs), cylindrical in shape, each connected to the main compartment through a truss.

The full shape of the geometric model is depicted in Fig. 2.10 as a three-dimensional image, labeled with the reference system used, and in Fig. 2.11 as a drawing with dimensions.

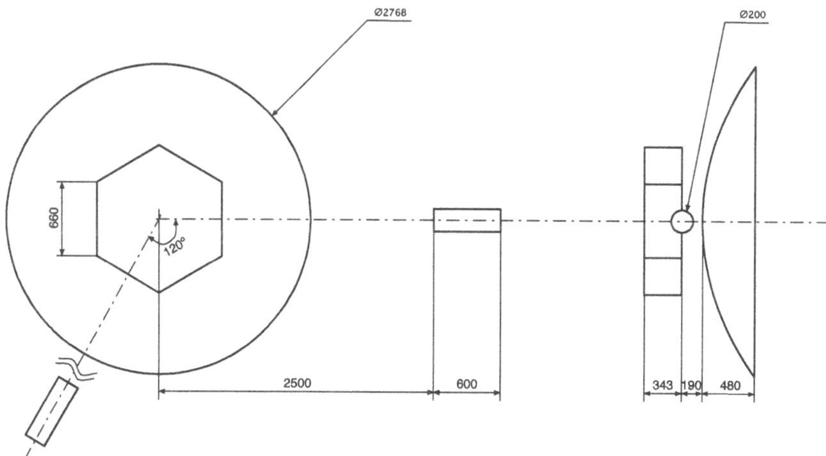

Fig. 2.11 Schematics of the Pioneer geometric model used in this study, with relevant dimensions (in mm); second RTG truss is not represented to scale. Lateral view indicates the relative position of the RTGs, box compartment and the gap between the latter and the high-gain antenna

This model simplifies the surface features and minor details of the spacecraft. This approximation has been tested through specific test-cases (*cf.* Sect. 2.3.3) which show that the effect on the final result is under 5 % and can be safely ignored for the purposes of this study.

2.4.2 Order of Magnitude Analysis

Before undertaking a more rigorous numerical computation, it is useful to perform a preliminary order of magnitude analysis. This allows one to obtain a concrete figure of merit for the overall acceleration arising from thermal effects, which can be compared with the $a_{\text{Pioneer}} \sim 10^{-9}$ m/s^2 scale of the Pioneer anomaly.

From the spacecraft specifications, one has a total mass $m_{\text{Pioneer}} \sim 230$ kg, and separate RTG and equipment compartment powers $W_{\text{RTG}} \sim 2$ kW and $W_{\text{equip}} \sim 100$ W, respectively. As already discussed, the integration of the emissions of the RTG and instrument compartment indicate the proportion of emitted power that is effectively converted into thrust. If we assume uniform temperature and emissivity in the RTGs and equipment compartment, we obtain

$$\begin{aligned} F_{\text{RTG}} &\sim 2 \times 10^{-2} \frac{W_{\text{RTG}}}{c}, \\ F_{\text{sides}} &\sim 10^{-1} \frac{W_{\text{equip}}}{c}, \\ F_{\text{front}} &\sim 2 \times 10^{-1} \frac{W_{\text{equip}}}{c}, \end{aligned} \tag{2.43}$$

where F_{RTG}, F_{sides} and F_{front} denote the contributions from the RTG, equipment side walls and front wall, respectively, and c is the speed of light.

Dividing by the mass, one can easily estimate the acceleration of the spacecraft due to the thermal effects arising from the power dissipation of the RTGs and equipment compartment

$$\begin{aligned} a_{\text{RTG}} &\sim 2 \times 10^{-2} \frac{W_{\text{RTG}}}{m_{\text{Pioneer}} c} \sim 6 \times 10^{-10} \text{ m/s}^2, \\ a_{\text{equip}} &\sim 3 \times 10^{-1} \frac{W_{\text{equip}}}{m_{\text{Pioneer}} c} \sim 4.4 \times 10^{-10} \text{ m/s}^2. \end{aligned} \tag{2.44}$$

This clearly indicates that both contributions are relevant to account for the reported anomalous acceleration. Furthermore, it also shows that the RTGs and the instrument compartment yield thermal effects of similar magnitudes, so that one cannot focus solely on one of these sources when modelling the spacecraft This had already been revealed by the analysis in Ref. [28].

2.4.3 Heat Conduction

Aside from radiative heat transfer, some conduction is expected between the structurally connected components of the spacecraft, thus affecting the distribution of radiated energy. Naturally, this will tend to warm the colder elements and cool the hotter ones, until a new equilibrium situation is achieved (more rigorously, a quasi-equilibrium, since the timescale of radiation and conduction is much smaller than that of the decreasing overall power output).

Thus, one should inspect the impact of this heat transfer in the main structural elements, namely between the RTGs and the main compartment and the main compartment and the high gain antenna. Conduction is expected to be larger between the RTGs and the main compartment: assuming that the latter is at about $0\,^{\circ}\mathrm{C}$ (a worst case scenario, as it is warmed by the electronics to $\sim 10\,^{\circ}\mathrm{C}$), while the RTGs are at $\sim 150\,^{\circ}\mathrm{C}$, a temperature gradient of approximately $60\ \mathrm{K/m}$ is obtained. The total cross-section of the three small diameter rods composing each truss is estimated to be of the order of $10^{-4}\ \mathrm{m}^2$; these are made of aluminium, with a conductivity of approximately $240\ \mathrm{W/(m\,K)}$. Using these figures, a total conducted power of the order of 1 W (up to 4 W in more conservative estimates) is obtained [22].

This is two orders of magnitude below the power of the main compartment and three orders of magnitude below the RTG power and can therefore be neglected. Since the temperature gradient between the main compartment to the antenna is much smaller than the one considered above, the related heat conduction is well below 1 W. Thus, one may safely disregard the contribution arising from conduction when computing the distribution of heat radiation.

2.4.4 Thermal Radiation Model

To begin the discussion surrounding the building of the thermal radiation model, it is important to highlight that this task is considerably simplified by the fact that the Pioneer probes are spin-stabilised. This implies that the effect of all emissions normal to the z axis are cancelled out after each complete revolution of the spacecraft, leaving only the contribution along the antenna's axis.

Using the point-like source method, we can now compute the contribution of the individual components listed in Table 2.8. This is achieved by integrating Eq. (2.39) in three successive steps. First, the emitted radiation field given by Eq. (2.35) is integrated along a closed surface, yielding the first-order effect due to the emissions. Afterwards, the same radiation field is integrated along the illuminated surfaces, in order to subtract the shadow effect. Finally, the reflected radiation vector-field, given by Eqs. (2.41) and (2.42), is integrated along closed surfaces, adding the contribution from reflection.

This process allows us to obtain the values for the force in terms of the emitted powers and reflection coefficients. As pointed out before, the results that follow

Table 2.8 Labelling of the considered contribution to the Pioneer 10 and 11 thermal acceleration

Emitting surface	Reflecting surface	Label
RTG	Side of main compartment	F_{12}
RTG	High-gain antenna dish	F_{15}
Side of main compartment	High-gain antenna dish	F_{25}
Front of main compartment	None	F_3
Back of main compartment	High-gain antenna dish	F_{45}

are only presented along the axis of the main antenna, since all radial components cancel-out. A positive figure indicates a sunward force.

The two main contributors to thermal emissions aboard Pioneer 10 and 11 are the RTGs, where the main power source of the spacecraft is located, and the main equipment compartment, where most of the generated electrical power is consumed. Emissions from the RTGs illuminate both the high-gain antenna and the lateral walls of the main equipment compartment. Radiation originating in the main compartment only significantly illuminates the main antenna, since the RTGs are too small and distant to yield any significant contribution from absorption or reflection.

The relevant contributions for this analysis are summarised in Table 2.8, with the labels used throughout the rest of this section. The complete set of radiation sources used in the model is listed in Table 2.9 and are described in the following paragraphs.

Table 2.9 Position and direction of the Lambertian sources used to model each emitting surface of the Pioneer spacecraft model

Emitting surface	Source	Position (m)	Surface normal (m)
Front wall (index 4)	1	$(0, 0, -0.343)$	$(0, 0, -1)$
Lateral wall (only one modelled)	1	$(0.572, 0.2475, -0.172)$	$(1, 0, 0)$
	2	$(0.572, 0.0825, -0.172)$	$(1, 0, 0)$
	3	$(0.572, -0.0825, -0.172)$	$(1, 0, 0)$
	4	$(0.572, -0.2475, -0.172)$	$(1, 0, 0)$
RTG (only one modelled)	1	$(2.5, 0, 0)$	$(-1, 0, 0)$
	2	$(3.1, 0, 0)$	$(1, 0, 0)$
Back wall	1	$(0.381, 0, 0)$	$(0, 0, 1)$
	2	$(0.191, 0.33, 0)$	$(0, 0, 1)$
	3	$(-0.193, 0, 33)$	$(0, 0, 1)$
	4	$(-0.381, 0, 0)$	$(0, 0, 1)$
	5	$(-0.191, -0.33, 0)$	$(0, 0, 1)$
	6	$(0.191, -0.33, 0)$	$(0, 0, 1)$

The front wall of the main equipment compartment (facing away from the sun and equipped with the heat-dissipating louvers) emits radiation directly into space, not illuminating any other surface. Furthermore, this surface is orthogonal to the spacecraft's spin axis. It can then be modelled by simply resorting to a single radiation source, without any impact on the accuracy of the final result.

The emitted radiation field is obtained by replacing the position of the radiation source and surface normal direction in Eq. (2.35). The force exerted by the radiation field on the emitting surface is obtained by integrating Eq. (2.39) along a closed surface. The resulting force exerted by the emitted radiation on the surface along the z axis is, as expected, given by

$$F_3 = \frac{2}{3}\frac{W_{\text{front}}}{c}, \tag{2.45}$$

where W_{front} is the emitted power.

The side walls of this compartment are each modelled by four Lambertian sources, as depicted in Fig. 2.12. The test cases and a previously performed convergence analysis showed that this is sufficient to reach the necessary accuracy [21, 22].

The radiation coming from the lateral walls of the main equipment compartment illuminates the high-gain antenna. Due to the symmetry of the problem, and neglecting the interaction with the small far RTGs, it is only necessary to model one of the six walls. The set of Lambertian sources used for one of these walls is indicated in Table 2.9. The z component of the radiation field force on the emitting surface vanishes, as the emitting surface is perpendicular to the z-axis.

Using Eq. (2.39), but taking the integral over the illuminated portion of the antenna dish, we obtain the force exerted on the illuminated surface, which accounts for the shadow effect. This gives a z component of $-0.0738(W_{\text{sides}}/c)$, where W_{sides} is the power emitted from the lateral walls, to be subtracted from the total force of the emitted radiation.

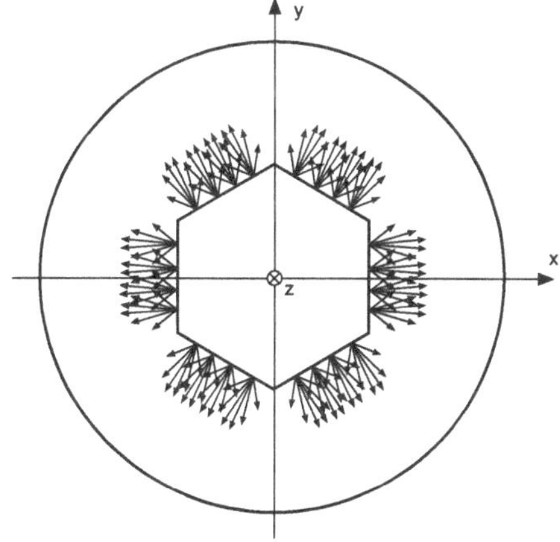

Fig. 2.12 Schematics of the configuration of Lambertian sources used to model the lateral walls of Pioneer's main equipment compartment

In what concerns the computation of diffusive reflection, Eq. (2.41) allows for the computation of the reflected Poynting vector-field $\mathbf{S}_{rd}(\mathbf{x}, \mathbf{x}')$ due to the emitted radiation field $\mathbf{S}(\mathbf{x}')$, where $\mathbf{x}'$ is a point in the reflecting surface. The reflected radiation field is given for each point in the reflecting surface. Consequently, it must be integrated first over the reflecting surface, conveniently parameterised, giving the resulting reflected radiation field, and then through Eq. (2.39) over a closed surface, in order to compute the force resulting from the reflected radiation. The procedure for specular reflection is analogous, except that Eq. (2.42) should instead be used to obtain the reflected radiation field.

Integrating the vector field representing radiation from the lateral walls of the main compartment reflecting on the high-gain antenna, we obtain a force result of $0.0537k_{d,ant}(W_{sides}/c)$ for the diffusive component and $0.0089k_{s,ant}(W_{sides}/c)$ for the specular component, where $k_{d,ant}$ and $k_{s,ant}$ are the diffusive and specular reflection coefficients of the main antenna, respectively.

The result for this contribution is given by adding the emitted radiation force (zero), the shadow effect and both components of reflection, leading to

$$F_{25} = \frac{W_{sides}}{c}(0.0738 + 0.0537k_{d,ant} + 0.0089k_{s,ant}), \tag{2.46}$$

where W_{sides} is the power emitted by the lateral walls of the main compartment, and $k_{d,ant}$ and $k_{s,ant}$ are the diffusive and specular reflection coefficients of the main antenna, respectively.

The back wall of the main equipment compartment faces the back of high-gain antenna. The radiation from this wall will, at a first iteration, reflect off the antenna and add a contribution to the force in the direction of the sun, as depicted in Fig. 2.13. This back wall was modelled using a set of six Lambertian sources evenly distributed in the hexagonal shape.

Initially, it was expected that the contribution from radiation emitted from the back wall of the main compartment and reflecting in the space between this compartment and the antenna dish would be small.

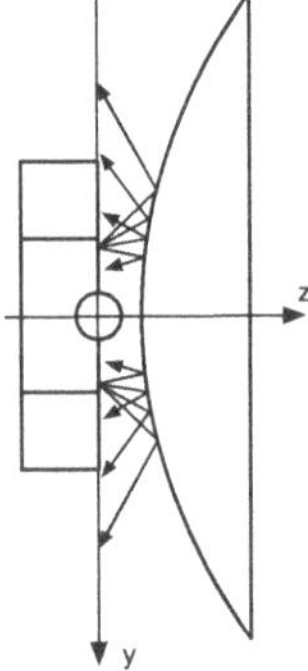

Fig. 2.13 Schematics of the configuration of Lambertian sources used to model the back wall of the main equipment compartment and the first reflection on the main antenna dish

In order to verify this assumption, a computation was made using the method described above. The results ultimately show that this contribution cannot be discarded after all, as it may be relevant in the final result. Considering one reflection from the antenna dish, the result in terms of the emitted power from the back wall of the main compartment W_{back}, by

$$F_{45} = \frac{W_{\text{back}}}{c}\left(-\frac{2}{3} + 0.5872 + 0.5040k_{\text{d,ant}} + 0.3479k_{\text{s,ant}}\right). \tag{2.47}$$

In the preceding equation, $-\frac{2}{3}\frac{W_{\text{back}}}{c}$ is the contribution from the emitted radiation and $0.5872\frac{W_4}{c}$ is the effect of the antenna's shadow. The remaining terms are the reflective contributions.

Finally, the RTGs can be easily and effectively modelled by two Lambertian sources, one at each base of the cylinder, as shown in Fig. 2.14. The emissions from the lateral walls need not to be considered since the radiation field has cylindrical symmetry and does not illuminate any surface. Also, radiation from the outward facing base radiates directly into space in a radial direction and its time averaged contribution is null. This leaves only the radiation emitted towards the centre of the spacecraft, that is reflected by both the high-gain antenna and the main equipment compartment.

Only one RTG needs to be modelled, since the effect of the components normal to the z axis will be cancelled-out at each revolution of the spacecraft.

Using the same procedure, the force generated by the RTG emissions is thus given in terms of the power emitted from the RTG bases facing the centre of the spacecraft W_{RTGb}. The force resulting from reflections on the antenna is given by

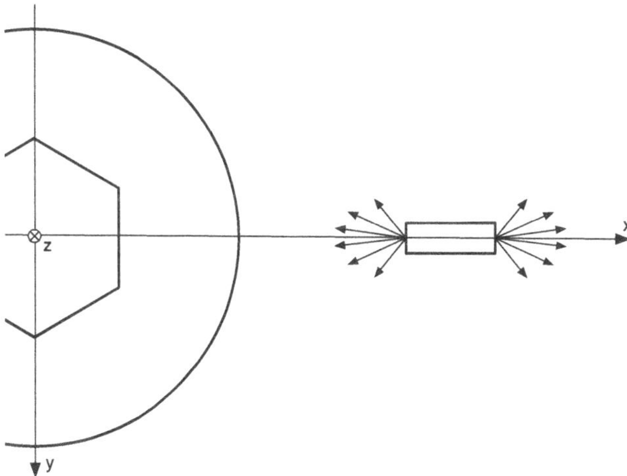

Fig. 2.14 Schematics of the two Lambertian sources used to model each RTG

$$F_{15} = \frac{W_{\mathrm{RTGb}}}{c}(0.0283 + 0.0478 k_{\mathrm{d,ant}} + 0.0502 k_{\mathrm{s,ant}}), \tag{2.48}$$

and the contribution from reflections on the lateral surfaces of the main equipment compartment is

$$F_{12} = \frac{W_{\mathrm{RTGb}}}{c}(-0.0016 + 0.0013 k_{\mathrm{s,sides}}), \tag{2.49}$$

where $k_{\mathrm{d,ant}}$, $k_{\mathrm{s,ant}}$, $k_{\mathrm{d,sides}}$ and $k_{\mathrm{s,sides}}$ are the respective reflection coefficients.

These force computations allow for obtaining, once the respective powers and reflection coefficients are inserted, of the acceleration due to thermal dissipation mechanisms is given by

$$a_{\mathrm{Pioneer}} = \frac{F_{12} + F_{15} + F_{25} + F_3 + F_{45}}{m_{\mathrm{Pioneer}}}, \tag{2.50}$$

where the spacecraft's mass is taken to be approximately $m_{\mathrm{Pioneer}} = 230$ kg. This figure considers a total launch mass of 259 kg, including 36 kg of hydrazine propellant that was partially consumed in the early stages of the mission [5]. Note that this is an approximate figure, since the actual masses for the Pioneer 10 and 11 would be slightly different due to different fuel consumptions along their respective missions.

2.4.5 Power Supply

The available onboard power was chosen as the independent variable in the computation of the thermally induced acceleration. This choice is justified with the fact that the available power is reasonably well known. Indeed, it is one of the few parameters with consistent data available throughout the operational life of the probes.

All the power on board the Pioneer probes comes from the two plutonium RTGs. It is thus easy to compute the total power available, considering the 87.74 year half-life of plutonium-238. According to Ref. [5], the total thermal power of the RTGs at launch was 2580 W. Consequently, its evolution with time should be given by

$$W_{\mathrm{tot}} = 2580 \exp\left(-\frac{t \ln 2}{87.72}\right) \text{ W}, \tag{2.51}$$

with t being the time in years after launch.

The electrical power is generated by a set of thermocouples located in the RTGs. Most of this power is consumed by the various systems located in the main equipment compartment, except for a small fraction used by the radio signal. Knowing the electrical power consumption, the remaining unused power is mostly dissipated at the RTGs themselves, through suitably designed radiating fins. Thus, it is reasonable to assume that the total available power is divided into electrical power used in equipment located in the main compartment and the remaining thermal power dissipated at the RTGs.

A good measurement of the electrical power is available through telemetry data. At launch, 120 W of electrical power were being used in the main equipment compartment plus around 20 W for the radio transmission to Earth, leaving 2420 W of thermal power in the RTGs. It is also known from telemetry data that the electrical power decayed at a faster rate than thermal power, with its half-life being around 24 years [28]. This would lead to an approximate time evolution of the electrical power in the equipment compartment given by

$$W_{\rm equip} \approx 120 \exp\left(-\frac{t \ln 2}{24}\right) \text{ W}. \tag{2.52}$$

The baseline scenarios established in the following section bear the above considerations in mind and accounts for the power values extracted from the available telemetry data for the latest stages of the mission. Specifically, the reading for the $t = 26$ years after launch. In a second stage of this study, the time evolution is taken into account, according to the reasoning developed in this section.

2.5 Results and Discussion

2.5.1 *Baseline Results*

In this section, a set of five scenarios is considered, while keeping the total power as $W_{\rm tot} = 2100$ W and the electrical power as $W_{\rm equip} = 56$ W, leaving RTG thermal power at $W_{\rm RTG} = 2024$ W (assuming the power of the radio beam is still 20 W).

The simplest possibility to consider is that each component of the spacecraft has a uniform temperature. In this case, the thermal power distribution is $W_{\rm front} = W_{\rm back} = 17.50$ W, $W_{\rm lat} = 21.00$ W and $W_{\rm RTGb} = 143.86$ W, leading to a thermally induced acceleration of

$$a_{\rm th,Sc1} = 2.27 \times 10^{-10} \text{ m/s}^2. \tag{2.53}$$

The second scenario assumes that the front wall, which is equipped with the louvers, is responsible for $W_{\rm front} = 40$ W (that is, 70 % of 56 W) of emission, maintaining the remaining walls with uniform emission. This scenario is motivated by the essential feature of the louvers being located in the front wall of the main equipment compartment. They were designed to act as a temperature controlling element, closing or opening through the action of a bi-metallic spring. Still, even when closed, the louvers are not covered by the Multi-Layer Insulation (MLI) which shields the equipment compartment. It is then reasonable to assume that, regardless of their position, the louvers radiate a large share of the equipment power. A similar argument is presented in Ref. [10]. This leaves the lateral walls with $W_{\rm sides} = 8.73$ W and the back wall with $W_{\mathrm{back}} = 7.27$ W. The thermally induced acceleration in these conditions is

$$a_{\mathrm{th,Sc2}} = 4.43 \times 10^{-10}\ \mathrm{m/s^2}. \tag{2.54}$$

In Scenario 3, one includes the contribution from reflections. The simplest way to achieve this is to include only the diffusive component. We consider a diffusive reflection coefficient of $k_{\mathrm{d,ant}} = 0.8$, which would be a typical value for aluminium, used in the antenna dish. This yields a result for the acceleration of

$$a_{\mathrm{th,Sc3}} = 5.71 \times 10^{-10}\ \mathrm{m/s^2}. \tag{2.55}$$

Scenario 4 is a variation that considers the fact that illuminated surface of the high-gain antenna is made of bare aluminium. This will make reflection from it mainly diffusive, but with a small specular highlight, as is typical of any unpolished flat metallic surface. We thus consider a reflection from the antenna dish that maintains a total reflection coefficient of 80 %, but divided in diffusive and specular components, respectively, $k_{\mathrm{d,ant}} = 0.6$ and $k_{\mathrm{s,ant}} = 0.2$. Furthermore, we assume a specular reflection from the MLI covering the main equipment compartment of $k_{\mathrm{s,sides}} = 0.4$. The result from this scenario is not significantly different from Scenario 3, yielding

$$a_{\mathrm{th,Sc4}} = 5.69 \times 10^{-10}\ \mathrm{m/s^2}. \tag{2.56}$$

In Scenario 5, in order to obtain an upper bound for the static baseline, one assumes that all the emissions of the main equipment compartment come from the louvers and a 10 % higher power from the RTG base, that is, $W_{\mathrm{front}} = 56$ W, $W_{\mathrm{back}} = W_{\mathrm{sides}} = 0$ W and $W_{\mathrm{RTGb}} = 158.24$ W. Maintaining $k_{\mathrm{d,ant}} = 0.8$, as in Scenario 3, the upper bound for the thermal acceleration in the late stage of the mission is bound to be

$$a_{\mathrm{th,Sc5}} = 6.71 \times 10^{-10}\ \mathrm{m/s^2}. \tag{2.57}$$

The hypotheses and results of all the analysed scenarios are summarised in Tables 2.10 and 2.11.

Table 2.10 Summary of the scenarios used to obtain baseline results for the Pioneer 10 and 11 thermal analysis

Scenario	Description
1	Uniform temperature, no reflection
2	Extra heat from louvers, no reflection
3	Scenario 2 with diffusive reflection
4	Scenario 2 with diffusive and specular reflection
5	Upper bound on all contributions

Table 2.11 Summary of results from baseline scenarios for the Pioneer 10 and 11 thermal analysis

Scenario	W_{RTGb} (W)	W_{front} (W)	W_{sides} (W)	W_{back} (W)	$k_{d,ant}$	$k_{s,ant}$	$k_{s,sides}$	$a_{Pioneer}$ (10^{-10} m/s^2)
1	143.86	17.5	21	17.5	0	0	0	2.27
2	143.86	40	8.73	7.27	0	0	0	4.43
3	143.86	40	8.73	7.27	0.8	0	0	5.71
4	143.86	40	8.73	7.27	0.6	0.2	0.4	5.69
5	158.24	56	0	0	0.8	0	0.4	6.71

W_{lb} is the power emitted from the base of the two RTGs; W_{front}, W_{lat} and W_{back} are, respectively, the emitted powers from the front, lateral and back walls of the main equipment compartment; $k_{d,ant}$ and $k_{s,ant}$ are the diffusive and specular reflection coefficients of the high-gain antenna back; $k_{s,sides}$ is the diffusive reflection coefficient of the lateral wall of the main equipment compartment, and $a_{Pioneer}$ is the resulting thermal acceleration along the rotation axis of the probe

With these baseline scenarios, we proceed with a static parametric analysis of the involved parameters in order to obtain a result and an error bar for the these static figures.

2.5.2 Parametric Analysis

As outlined above, the next step is to perform a static parametric analysis aiming to establish an estimate for the thermal acceleration at an instant 26 years after launch. The analysis is performed using a classic Monte-Carlo method, where a probability distribution is assigned to each variable and random values are then generated. A distribution of the final result (i.e. the acceleration) is then obtained.

The parameters that come into play in this setup are the power emitted from each surface, W_{RTGb}, W_{front}, W_{sides}, W_{back}, and the reflection coefficients $k_{d,ant}$, $k_{s,ant}$ and $k_{s,lat}$. The assertions made in the previous section regarding total power and its distribution between the RTGs and the equipment compartment remain valid.

A quick analysis of Table 2.11 allows us to draw some qualitative conclusions. For example, the amount of power emitted from the front wall W_{front} has a decisive influence in the final result. In contrast, the relevance of the specular reflection coefficient of the lateral wall $k_{s,lat}$ is almost negligible.

For the static analysis at $t = 26$ years, Scenario 4 is taken as a reference, since it is the one more solidly based on physical arguments.

The power emitted by the RTG bases facing the main compartment W_{RTGb} is generated from a normal distribution with the mean value of 143.86 W and a standard deviation of 25 % of this value. This allows for significantly larger deviation than the considered in the top-bound scenario (Scenario 5), which had only a 10 % increase in the power of this surface. The purpose is to account for unanticipated anisotropies in the temperature distribution of the RTGs.

In the case of the main equipment compartment, the focus is on the power emitted by the louvers located in the front wall. The selected distribution for the parameter W_{front} is also normal, with the mean value at 40 W (also corresponding to Scenario 4). We set the standard deviation at 7.5 W, so that the 95 % probability interval (2σ) for the value of W_{front} is below the top figure of 56 W, which corresponds to the totality of the equipment power being dissipated in the front wall. For the remaining surfaces of the equipment compartment, the power is computed at each instance so that the total power of the equipment is kept at 56 W.

Concerning the reflection coefficients for the antenna, we use uniform distributions in the intervals [0.6, 0.8] for $k_{\text{d,ant}}$ and [0, 0.2] for $k_{\text{s,ant}}$, while imposing the condition $k_{\text{d,ant}} + k_{\text{s,ant}} = 0.8$, since this is a typical value for aluminium in infrared wavelengths. We also expect the specular component to be small, since the surface is not polished. Furthermore, if we allow for the possibility of surface degradation with time during the mission, the specular component would suffer a progressive reduction in favour of the diffusive component, a possibility that this analysis takes into account.

We performed 10^4 Monte Carlo iterations, which easily ensures the convergence of the result. The thermal acceleration estimate yielded by the simulation for an instant 26 years after launch, with a 95 % probability level (2σ), is

$$a_{\text{th}}(26) = (5.8 \pm 1.3) \times 10^{-10} \text{ m/s}^2. \tag{2.58}$$

This result is extracted from the approximately normal distribution shown in Fig. 2.15. The conformity of the results to a normal distribution was confirmed by a Shapiro-Wilk normalcy test, and can also be qualitatively evaluated in the graph.

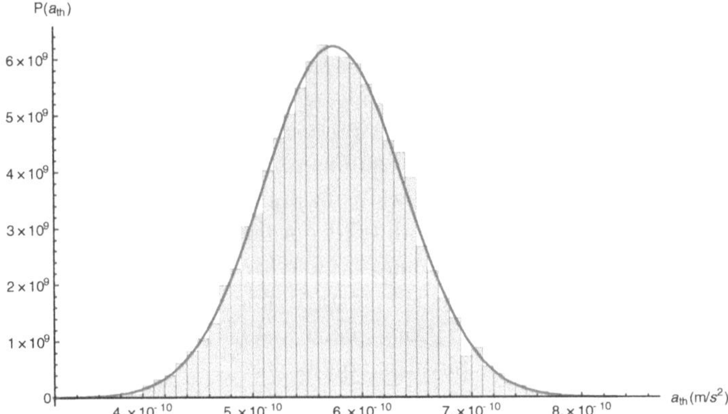

Fig. 2.15 Histogram for the probability density distribution resulting from the Monte Carlo simulation with 10,000 iterations for the thermal acceleration of the spacecraft at $t = 26$ years after launch. A Normal distribution with the same mean value and standard deviation is superimposed

These results account for between 44 and 96 % of the initially reported value $a_{\text{Pio}} = (8.74 \pm 1.33) \times 10^{-10}$ m/s^2, which, we recall, was obtained under the hypothesis of a constant acceleration. This alone, already gives a strong indication of the preponderant contribution of thermal effects to the Pioneer anomaly. However, we must proceed to obtain a time signature that can be compared with the more recent studies of the anomaly [20].

2.5.3 Time Evolution

Since thermal effects depend on the radioactive decay of the plutonium and the electrical power produced, they change with time. The final step is thus to perform an analysis of the expected time evolution of the thermal acceleration affecting the Pioneer 10 and 11 spacecraft.

An immediate estimate can be obtained by extrapolating the static results with the available time evolution of electric power, using Eqs. (2.51) and (2.52). This extrapolation, however, does not account for the possibility that some parameters may change with time, namely, the power distribution throughout the different surfaces or their reflection coefficients. This could be accounted by a simulation of the full span of the missions (i.e. a large number of consecutive simulations), with a specific prescription for the variability of these parameters.

Such task would prove too lengthy, and no significant physical insight would be gained. Hence, we have preferred a somewhat simpler approach: for a better grasp of the possibility discussed above, we apply the Monte-Carlo static analysis to only two earlier moments of the mission. Each simulation produces a central value, with top and lower bounds; these are then fitted to an exponential trend, thus obtaining an estimate of the time evolution of the thermally induced acceleration.

The selected instant for the earliest static analysis was at $t = 8$ years after launch, corresponding to the 1980 values for the Pioneer 10. This corresponds to the time at which the effect of the solar radiation pressure dropped below 5×10^{-10} m/s^2 [5].

This analysis is made in a similar fashion as the one presented in the previous section, but using the 1980 available power values as a base for the choice of the distributions. The thermal acceleration is, in this case,

$$a_{\text{th}}(8) = (8.9 \pm 2) \times 10^{-10}\ \text{m/s}^2, \tag{2.59}$$

corresponding to the same 95 % probability level in the approximately normal distribution in Fig. 2.16.

The values obtained here for this earlier stage of the mission bear a close match to those of the assumed constant anomalous acceleration.

The third static analysis was performed at a time $t = 17$ years, halfway between the other two. The estimate in this case is, for a 95 % probability,

$$a_{\text{th}}(t = 17) = (7.1 \pm 1.6) \times 10^{-10}\ \text{m/s}^2. \tag{2.60}$$

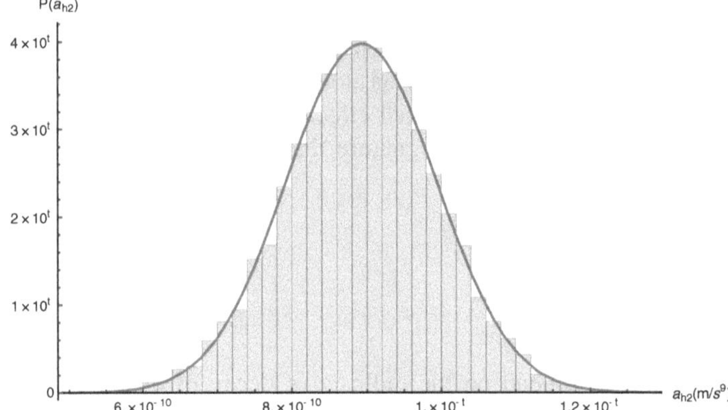

Fig. 2.16 Histogram for the probability density distribution resulting from the Monte Carlo simulation with 10,000 iterations for the thermal acceleration of the spacecraft at $t = 8$ years after launch. A Normal distribution with the same mean value and standard deviation is superimposed

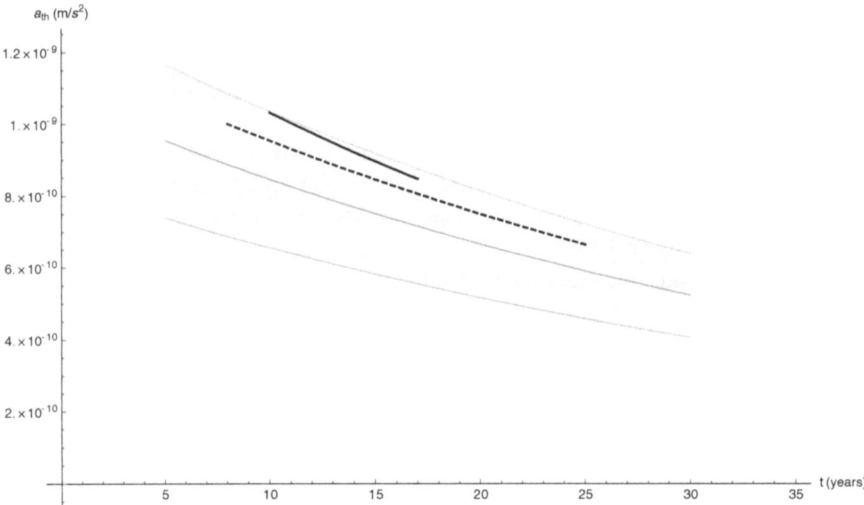

Fig. 2.17 Results for the time evolution of the thermal acceleration on the Pioneer spacecraft compared with results based on the latest data analysis of the anomalous acceleration. The *shaded area* correspond to a 95 % probability for the thermal acceleration in the time evolution analysis. For comparison, the *black dashed* and *solid lines* are based on results from the data analysis in Ref. [20]

Using the three static estimates presented above, it is now possible to produce a time evolution based on a fit to an exponential decay. This is performed for the mean value, top-bound and lower-bound of the acceleration, always based on a 2σ interval.

The curve fit for the mean, upper and lower values of the thermal acceleration reads

$$a_{\text{th}} = [(1.07 \pm 0.24) \times 10^{-9}]e^{-0.0240t} \text{ m/s}^2, \quad (2.61)$$

with t giving the time after launch in years. The time constant was averaged between the three fits, although they are all within 1 % of each other.

The time evolution resulting from any of these scenarios corresponds to a decay with a half-life of approximately 29 years, related to the nuclear decay of the plutonium in the RTGs and the faster decay rate of the electrical power, already discussed in Sect. 2.4.5. The graphic representation of the band of values predicted by our model is shown in Fig. 2.17 (shaded region) and compared with the values indicated by the exponential fit results for the anomalous accelerations of Pioneer 10 and 11 in Ref. [20] (blach dashed and solid lines, respectively).

2.6 Conclusion

In this chapter, an entirely new method to compute spacecraft thermally induced accelerations has been presented and tested. This method was specifically developed to tackle problems where engineering and construction details are scattered and not easily available. Despite its apparent simplicity, the battery of test cases to which it was subjected clearly shows its reliability by ensuring that the errors involved in the simplification of the radiation modelling are well below 10 %.

The parametric analysis that follows the radiation modelling further enhances the trustworthiness of the results. Assumptions are made based on available information, minimising the amount of guessing involved. The width of the resulting interval, for a set probability level, reflect the accuracy of the input data available to construct the model. The more information there is, the narrower the interval will be.

The application of the point-like source method to the Pioneer anomaly produced the first reliable accounting of the thermally induced acceleration on the Pioneer 10 and 11. The obtained acceleration values, as well as the temporal signature, are compatible with the latest descriptions of the detected anomalous acceleration. After our results, it became highly unlikely that the Pioneer anomaly would be explained by anything other than the forces arising from anisotropic thermal radiation.

The confirmation of our results by the ZARM [24] and JPL [25], both using finite-element models, gave the final piece of the puzzle that was missing before the Pioneer anomaly could finally be closed.

References

1. Pioneer Odyssey: Encounter with the Giant (2014), http://history.nasa.gov/SP-349/ch8.htm. Accessed 02 Nov 2014
2. NASA Solar System Exploration Multimedia Gallery (2014), http://solarsystem.nasa.gov/multimedia/display.cfm?Category=Spacecraft&IM_ID=16348. Accessed 02 Nov 2014

3. NASA Solar System Exploration Multimedia Gallery (2014), http://solarsystem.nasa.gov/multimedia/display.cfm?IM_ID=2903. Accessed 02 Nov 2014
4. The Pioneer Missions (2014), http://www.nasa.gov/centers/ames/missions/archive/pioneer.html. Accessed 02 Nov 2014
5. J. Anderson, P. Laing, E. Lau, A. Liu, M. Nieto, S.G. Turyshev, Study of the anomalous acceleration of Pioneer 10 and 11. Phys. Rev. D **65**(8), 082004 (2002). doi:10.1103/PhysRevD.65.082004
6. J. Anderson, P. Laing, E. Lau, A. Liu, M. Nieto, S.G. Turyshev, Indication, from Pioneer 10/11, Galileo, and Ulysses data, of an apparent anomalous, weak, long-range acceleration. Phys. Rev. Lett. **81**(14), 2858–2861 (1998). doi:10.1103/PhysRevLett.81.2858
7. E. Murphy, Prosaic explanation for the anomalous accelerations seen in distant spacecraft. Phys. Rev. Lett. **83**(9), 1999 (1890). doi:10.1103/PhysRevLett.83.1890
8. J. Katz, Comment on indication, from Pioneer 10/11, Galileo, and Ulysses data, of an apparent anomalous, weak, long-range acceleration. Phys. Rev. Lett. **83**(9), 1999 (1892). doi:10.1103/PhysRevLett.83.1892
9. O. Bertalmai, J. Páramos, A mission to test the Pioneer anomaly: estimating the main systematic effects. Int. J. Mod. Phys. D **16**, 1611 (2007). doi:10.1142/S0218271807011000
10. L. Scheffer, Conventional forces can explain the anomalous acceleration of Pioneer 10. Phys. Rev. D **67**(8), 084021 (2003). doi:10.1103/PhysRevD.67.084021
11. O. Bertolami, J. Páramos, The Pioneer anomaly in the context of the braneworld scenario. Class. Quantum Gravity **21**(13), 3309–3321 (2004). doi:10.1088/0264-9381/21/13/013
12. J.R. Brownstein, J.W. Moffat, Gravitational solution to the Pioneer 10/11 anomaly. Class. Quantum Gravity **23**(10), 3427–3436 (2006). doi:10.1088/0264-9381/23/10/013
13. M.-T. Jaekel, S. Reynaud, Post-Einsteinian tests of linearized gravitation. Class. Quantum Gravity **22**(11), 2135–2157 (2005). doi:10.1088/0264-9381/22/11/015
14. O. Bertolami, T. Harko, F.S.N. Lobo, Extra force in f(R) modified theories of gravity. Phys. Rev. D **75**(10), 104016 (2007). doi:10.1103/PhysRevD.75.104016
15. S.G. Turyshev, V.T. Toth, The Pioneer anomaly. Living Rev. Relativ. **13**, 4 (2010). doi:10.12942/lrr-2010-4
16. O. Bertolami, P. Vieira, Pioneer anomaly and the Kuiper Belt mass distribution. Class. Quantum Gravity **23**(14), 4625–4635 (2006). doi:10.1088/0264-9381/23/14/005
17. C.B. Markwardt, Independent Confirmation of the Pioneer 10 Anomalous Acceleration (2002), arXiv:gr-qc/0208046v1
18. A. Levy, B. Christophe, P. Bério, G. Métris, J.M. Courty, S. Reynaud, Pioneer 10 Doppler data analysis: disentangling periodic and secular anomalies. Adv. Space Res. **43**(10), 1538–1544 (2009). doi:10.1016/j.asr.2009.01.003
19. V. Toth, Independent analysis of the orbits of Pioneer 10 and 11. Int. J. Mod. Phys. D **18**(05), 717–741 (2009). doi:10.1142/S0218271809014728
20. S.G. Turyshev, V.T. Toth, J. Ellis, C.B. Markwardt, Support for temporally varying behavior of the Pioneer anomaly from the extended Pioneer 10 and 11 Doppler data sets. Phys. Rev. Lett. **107**(8) (2011). doi:10.1103/PhysRevLett.107.081103
21. O. Bertolami, F. Francisco, P.J.S. Gil, J. Páramos, Thermal analysis of the Pioneer anomaly: a method to estimate radiative momentum transfer. Phys. Rev. D **78**(10), 103001 (2008). doi:10.1103/PhysRevD.78.103001
22. O. Bertolami, F. Francisco, P.J.S. Gil, J. Páramos, Estimating radiative momentum transfer through a thermal analysis of the Pioneer anomaly. Space Sci. Rev. **151**(1–3), 75–91 (2010). doi:10.1007/s11214-009-9589-3
23. F. Francisco, O. Bertolami, P.J.S. Gil, J. Páramos, Modelling the reflective thermal contribution to the acceleration of the Pioneer spacecraft. Phys. Lett. B **711**(5), 337–346 (2012). doi:10.1016/j.physletb.2012.04.034
24. B. Rievers, C. Lämmerzahl, High precision thermal modeling of complex systems with application to the flyby and Pioneer anomaly. Ann. Phys **523**(6), 439–449 (2011). doi:10.1002/andp.201100081

25. S.G. Turyshev, V. Toth, G. Kinsella, S.-C. Lee, S.M. Lok, J. Ellis, Support for the thermal origin of the Pioneer anomaly. Phys. Rev. Lett. **108**(24), 241101 (2012). doi:10.1103/PhysRevLett.108.241101
26. O. Bertolami, F. Francisco, P.J.S. Gil, J. Páramos, Modeling the nongravitational acceleration during Cassini's gravitation experiments. Phys. Rev. D **90**(4), 042004 (2014). doi:10.1103/PhysRevD.90.042004
27. B.T. Phong, Illumination for computer generated pictures. Commun. ACM **18**(6), 311–317 (1975). doi:10.1145/360825.360839
28. V. Toth, S.G. Turyshev, Pioneer anomaly: evaluating newly recovered data. AIP Conf. Proc. **977**, 264–283 (2008). doi:10.1063/1.2902790

Chapter 3
Cassini Gravitational Experiments

3.1 General Background

The Cassini mission was launched on October 15th 1997. Its goal was to reach the Saturn system and study the planet and its natural satellites. Notably, the Cassini orbiter, depicted in Fig. 3.1 [1], dropped the Huygens atmospheric descent probe, which performed the first ever landing in the outer solar system in 2005 when it reached the surface of Titan [2].

Along the mission a set of experiments designed to test General Relativity (GR) were performed. One of these experiments was carried out from June 6th to July 7th 2002, while the probe was near solar conjunction with Earth, measuring the time delay of the radio signal to measure the γ parameter of the Parameterised Post-Newtonian (PPN) formalism.

In the PPN formalism, γ measures the amount of curvature produced per unit mass and its value in GR is $\gamma = 1$ [3–5]. The two main processes to test this parameter are through the deflection of light and through its time delay. The Cassini experimental setup uses the latter [6].

During solar conjunction, a radio beam travelling between the spacecraft and Earth passes near the Sun, experiencing a time delay due to its gravitational field, as depicted in Fig 3.2. The time delay in the round trip of a radio beam from the probe to Earth and back is given, in terms of the γ parameter by

$$\Delta t = 2(1+\gamma)\frac{GM_{\odot}}{c^3}\ln\left(\frac{4r_{gs}r_{sc}}{b^2}\right) \tag{3.1}$$

where G is the gravitational constant; $M_{\odot}$ is the Sun's mass; c is the speed of light; r_{gs} and r_{sc} are the distances from the Sun to the ground station and the spacecraft, respectively, and $b \ll r_{gs}, r_{sc}$ is the impact parameter, as shown in Fig. 3.2 [6, 7].

As both the spacecraft and Earth move in their orbits, this Δt will change with time, producing a gravitational frequency shift given by

F. Francisco, *Trajectory Anomalies in Interplanetary Spacecraft*, Springer Theses,
DOI 10.1007/978-3-319-18980-2_3

Fig. 3.1 Artist's impression of Cassini during its Saturn orbit insertion.]Artist's impression of Cassini during its Saturn orbit insertion [1]. The asymetric configuration of the RTGs and the umbrella like structures covering them are clearly visible in this picture

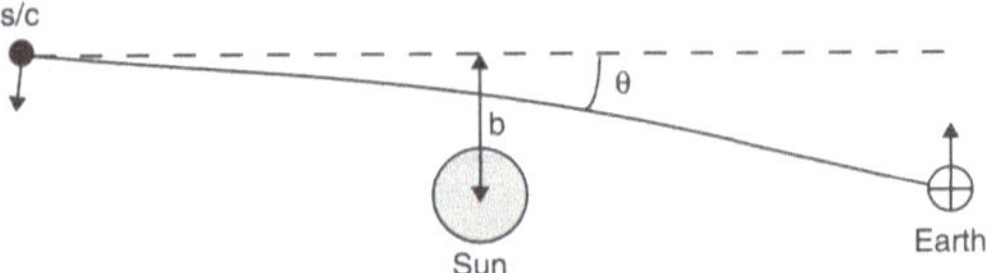

Fig. 3.2 Geometry of the trajectory of a radio beam travelling between a spacecraft and earth near solar conjunction. Taken from Ref. [6]

$$y_{gr} = \frac{d\Delta t}{dt} = -2(1+\gamma)\frac{GM_{\odot}}{c^3 b}\frac{db}{dt}. \tag{3.2}$$

The measurement of this observable, which produces the set of data represented in the graph in Fig 3.3, allows for the determination of γ [6, 7].

Prior to the Cassini experiments γ was constrained to within 0.001 of unity. The new results from the data harvested during this one month period allowed for constraining γ to within $(2.1 \pm 2.3) \times 10^{-5}$ of unity, the most accurate bound obtained so far [7].

During the solar conjunction experiment, the non-gravitational acceleration had to be filtered out as well as possible and, in particular, the significant contributions from solar radiation pressure and from anisotropic thermal emission arising from the probe itself. Due to the unavailability of any straightforward procedure to obtain the said thermal emission from a model of the spacecraft, data from Doppler measurements was used to estimate the component of the acceleration that is constant relative to the spacecraft orientation. The obtained values for the thermally generated

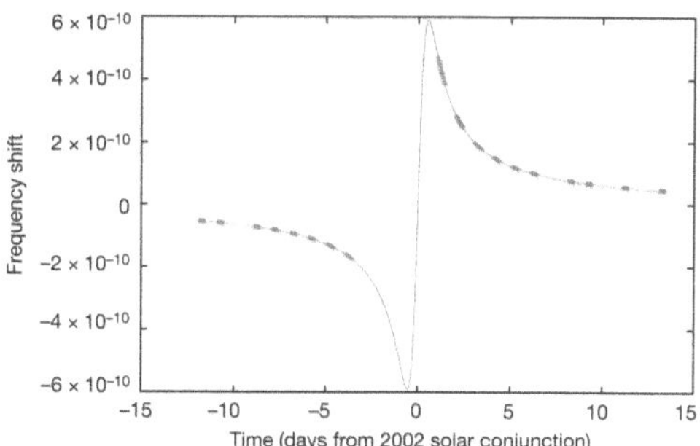

Fig. 3.3 Gravitational shift data from Cassini solar conjunction experiment. Taken from Ref. [7]

acceleration reveal that the largest component is aligned with the Earth-spacecraft axis and amounts to 3×10^{-9} m/s^2 towards the Earth. The other two components are smaller and measured orthogonally to the orbital plane and on the orbital plane, and are found to be about 4×10^{-10} and 1×10^{-10} m/s^2, respectively. These components, however, have large error estimates associated with their determination [7].

In this Chapter we consider the problem of obtaining the value of the thermally generated accelerations and directly respond to the stated difficulty in extracting them from a model of the spacecraft. The analysis is made using the point-like source method, presented in detail in Sect. 2.3.

This presented an ideal opportunity to put the method to the test in a whole new situation, strengthening the confidence of both method and results. It was, once again, shown that reliable results can be obtained by using the physical and computational framework developed for that problem. The flexibility and adaptability of the method is demonstrated, since some key features of this analysis are somewhat different from Pioneer's, notably the stabilisation method of the probes.

Ultimately, this work resulted in a value for the main component of the thermal acceleration that closely matches the estimate performed by Bertotti et. al., certifying the credibility of the gravitational experiment [8].

3.2 Cassini Thermal Model

3.2.1 Geometric Model

The first step in this analysis is to build a simplified geometric model of the spacecraft that retains only its main features. This procedure has been validated by a set of test cases performed previously in the analysis of the Pioneer space probes, which gave a good indication that the effect of smaller features does not significantly impact the overall determination of the thermal contribution to the acceleration and can be taken into account in the parametric analysis.

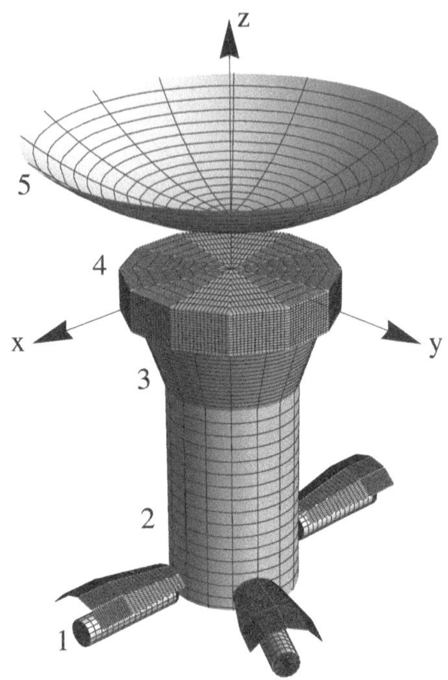

Fig. 3.4 Three-dimensional simplified geometric model of Cassini showing the configuration of the RTGs and their covering structures (*1*), the spacecraft body composed by a cylindrical lower module (*2*), a conical upper module (*3*), a prismatic main bus (*4*) and the parabolic high-gain antenna (*5*)

In what respects Cassini, this means the inclusion of the main antenna dish, the spacecraft body, the three Radioisotope Thermal Generators (RTGs) and respective covers. Cassini's main body is composed of a main bus shaped like a dodecagonal prism, an upper module with a conic shape and a cylindrical lower module. The three RTGs are attached to the lower module near to its bottom in an asymmetrical configuration (*cf.* Fig. 3.1). While two of the RTGs are in diametrically opposite positions, the third one is mounted at an angle of 120°. Each RTG is covered by an umbrella-like structure composed of eight flat surfaces, arranged as shown in Fig. 3.4.

Unlike Pioneer 10 and 11, Cassini is not spin stabilised. Instead, it uses an active three-axis stabilisation with spin-wheels and thrusters. Due to this fact, the off-axis components of the force are not canceled over time and have to be computed. In any case, judging from the probe's configuration, the component along the z-axis should still be dominant. Coincidentally, it is also the component for which there is more reliable data available for comparison.

3.2.2 Order of Magnitude Analysis

Before embarking on a systematic effort to model the thermal effects on Cassini, an order of magnitude analysis of the different contributions can provide valuable insight, particularly, by giving a first assessment onto the relative importance of the different contributions.

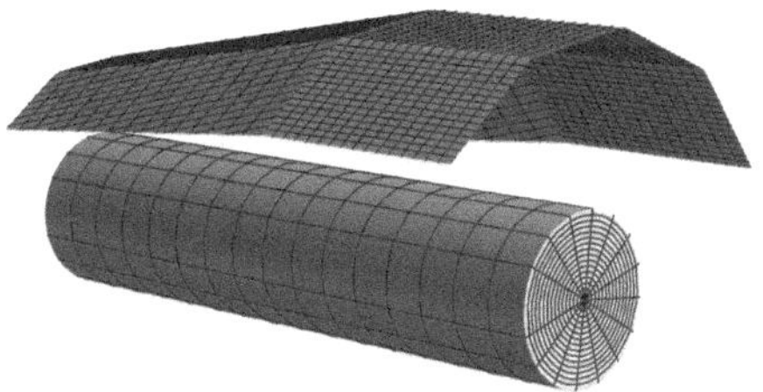

Fig. 3.5 Detail of the geometric model of the umbrella-like structure covering each RTG

To this end, it is enough to consider, for now, that the combined power of the three RTGs is on the order of 10 kW and the available electrical power for all the equipment is on the order of 1 kW.

The configuration of the RTGs, each covered with an umbrella-like structure, as shown in Fig. 3.5, ensures that a large fraction of the emitted thermal power is absorbed or reflected by those covers, surely leading to a significant contribution to the thermal force.

From the model of the RTG covers, we find that around 30 % of the power emitted by the RTGs, W_{RTG}, illuminates the umbrella-like shields. The shadow cast by these structures, means that the radiation emitted towards the main antenna no longer compensates the radiation propagating in the opposite direction on an otherwise cylindrical radiation field. It is then reasonable at this point to take this value and assume that 30 % of the emissions from the RTG are converted into a force in the positive z direction. With these assumptions, the order of magnitude of the force generated by the RTGs amounts to

$$F_{\text{RTG}} \sim 0.3\frac{W_{\text{RTG}}}{c} \sim 10^{-5}\,\text{N}. \tag{3.3}$$

Dividing by the spacecraft mass, which, for now, is assumed to be on the order of 4600 kg [7], we obtain the expected magnitude of the thermal acceleration generated by the RTGs

$$a_{\text{RTG}} \sim 2.2 \times 10^{-9}\,\text{m/s}^2. \tag{3.4}$$

When examining the spacecraft body, we can set an upper bound for its contribution, so that it can be compared with the estimates for the effect of the RTGs. To do so, one can assume that all the electrical power is dissipated through the bottom wall of the lower compartment. This scenario, although simplistic, maximises the effect of the thermal radiation from the equipment. Under these conditions, we get an upper bound on the force generated by the electrical equipment in the body of about

$$F_{\text{equip}} \lesssim \frac{2}{3}\frac{W_{\text{elec}}}{c} \sim 2.2 \times 10^{-6}\,\text{N}, \tag{3.5}$$

and, at most, an acceleration of

$$a_{\text{equip}} \lesssim 4.8 \times 10^{-10}\ \text{m/s}^2, \tag{3.6}$$

which is below the estimated effect of the RTGs by a factor of 5. We stress that this figure clearly overestimates the effects of thermal radiation from electrical power. Any detailed computation is likely to yield a much smaller figure.

This preliminary analysis allows us to conclude that the contribution from the RTGs is dominant in what concerns Cassini's thermal acceleration. The obtained order of magnitude also matches the one of the acceleration estimated from the Doppler data.

3.2.3 Thermal Radiation Model

Based on the results of the preceding section, our attention is focused on the contribution of the RTGs in the first place. We have already concluded that a very significant amount of the radiation emitted from the RTGs is illuminating their covers.

The geometric model of the illuminated surface is, in this case, quite realistic, as shown in Fig. 3.5, since the flat surfaces are easily modelled without need for any simplifcation. The main issue here is to obtain the correct distribution of radiation sources that effectively models the emissions of the RTGs.

In order to discern the sensitivity of the result, we built three different models for the RTG:

(i) four isotropic sources uniformly distributed along the centreline,
(ii) a cylindrical source along the centreline, and
(iii) 24 Lambertian sources distributed in four rings of six sources each along the lateral surface of the RTG.

The model with four isotropic sources somewhat underestimates the effect, with a deviation of around 20 % relative to the other two. This is due to the fact that it has a significant amount of radiation being emitted laterally in what would still be inside the volume of the RTG. Both other models closely reproduce the cylindrical radiation field, with the results for the power illuminating the RTG cover given by both models within 5 % of each other, which is well within the accuracy goals set for this work.

After analysing and comparing the results from these models, it was decided to use the cylindrical source configuration, since it is the one that models the radiation field in a more realistic manner and computationally lighter when compared to the 24 source model.

The first result to be obtained is the fraction of the power emitted that illuminates the covering structure. This figure comes in at 28.3 %, a part of which is absorbed and the remaining is reflected, depending on the optical properties of the inner surface of the RTG covers. The force computation is made leaving the diffusive and specular reflection coefficients as an open variable to be dealt with later on.

It should be recalled that Cassini has three RTGs positioned in an asymmetrical configuration. For that reason, at this point only the contribution of a single RTG is computed, for simplicity, the one aligned along the x-axis. The contributions of the other two can be obtained afterwards through a simple rotation matrix. Using the reflection modelling described in Sect. 2.3.4 and performing the numerical integration, we obtain the force resulting from the emissions of that single RTG

$$\mathbf{F}_{11} = \frac{W_{\mathrm{RTG}}}{c}\Big[(-0.0356 - 0.0204k_{\mathrm{d,umbr}} - 0.0466k_{\mathrm{s,umbr}})\mathbf{e}_x + (0.240 + 0.159k_{\mathrm{d,umbr}} + 0.193k_{\mathrm{s,umbr}})\mathbf{e}_z\Big], \tag{3.7}$$

where W_{RTG} is the power emitted by the RTG and $k_{\mathrm{d,umbr}}$ and $k_{\mathrm{s,umbr}}$ are the diffusive and specular reflection coefficients of the inner surface of the RTG cover, respectively. In order to obtain the total contribution from the 3 RTGs, we have to add this result to the contribution rotated by 120° and 180° around the spacecraft's z-axis. This and the other contributions to the force are labeled according to Table 3.1, using the spacecraft part numbering from Fig. 3.4.

Aside from the RTG contribution, the electric power consumed by the equipment in the spacecraft body also causes thermal acceleration. However, the order of magnitude analysis performed in Sect. 3.2.2 shows that, at most, it adds up to around 20 % of the contribution of the RTGs. Still, the effect of the top and bottom walls can be significant along the z-axis and deserves some effort in its determination.

When evaluating the emissions from the top wall of the spacecraft main bus, the main surface being illuminated is the back of the parabolic high-gain antenna. The emissions from this surface were modelled through a total of 12 Lambertian sources, each one placed at the centroid of each triangular segment of the dodecagon shaped surface.

Integrating along the antenna, we find that 61.1 % of the thermal power emitted from the top wall is hitting the antenna. Assuming that the power is evenly distributed along the surface, the radial components of the source cancels out, leaving only an axial contribution of

$$\mathbf{F}_{45} = \frac{W_{\mathrm{top}}}{c}\left(-\frac{2}{3} + 0.492 + 0.387k_{\mathrm{d,ant}} + 0.236k_{\mathrm{s,ant}}\right)\mathbf{e}_z, \tag{3.8}$$

where W_{top} is the power emitted from the top wall, $k_{\mathrm{d,ant}}$ is the diffusive reflection coefficient of the antenna and $k_{\mathrm{s,ant}}$ is its specular reflection coefficient.

Table 3.1 Labelling of the considered contributions to the Cassini thermal acceleration

Emitting surface	Reflecting surface	Label
RTGs	RTG covers	F_{11}
Main bus upper wall	High-gain antenna back wall	F_{45}
Lower module bottom wall	None	F_2

It should be noted that both the inner wall of the umbrellas and the lower surface of the antenna are modelled with a low shininess constant, $\alpha = 3$ (*cf.* Eq. 2.4), since these are unpolished surfaces.

Any amount of power emitted from the bottom wall yields a direct contribution to the acceleration along the z-axis, since it does not illuminate any other surface. Considering that it is a Lambertian emitter, if W_{bottom} is the power emitted from the bottom wall, then its contribution to the force is

$$\mathbf{F}_2 = \frac{2}{3}\frac{W_{\text{bottom}}}{c}\mathbf{e}_z. \tag{3.9}$$

Given the available information, there is no way to obtain any detailed distribution of the thermal emissions on the lateral walls of the main body of the spacecraft. In any case, such contribution should be very small, since the body has an approximately cylindrical shape and the multilayer insulation blanket tends to even out the temperatures, making the radial radiation field emanating from the spacecraft body approximately symmetric. For this reason, we keep our focus mainly on the acceleration component along the Earth-spacecraft axis, approximately coincident with Cassini's z axis, while attempting to get a rough estimate of the other component based entirely on the effect of the RTGs.

3.2.4 Power Supply

The amount of power available on board is of crucial importance for the outcome. Cassini is powered by a set of three large plutonium RTGs. At launch, these RTGs generated around 13 kW of total thermal power, from which 878 W of electrical power were produced. Since the plutonium decays with a half-life of 87.7 years, the total thermal power W_{Total} should decrease at the same rate.

Electrical power is generated by the RTGs trough a set of thermocouples. Due to the decay in the conversion efficiency, its decay is faster than the thermal power loss. This rate of decay can be fitted by an exponential law with a half-life of approximately 21.2 years [9]. Taking these combined effects into account, the time evolution of the electrical power is given by

$$W_{\text{elec}}(t) = 878e^{-\frac{t\ln 2}{87.7}}e^{-\frac{t\ln 2}{21.2}}\ \text{W} = 878e^{-\frac{t\ln 2}{17.1}}\ \text{W}, \tag{3.10}$$

with t in years, thus yielding a combined half-life of 17.1 years.

In order to maintain the overall spacecraft energy balance, we assume that the thermal power dissipated at the RTGs results from the difference between total thermal power and the electrical power generated, since the latter will be used to power the array of equipment carried in the spacecraft body,

$$W_{\text{RTG}}(t) = W_{\text{total}}(t) - W_{\text{elec}}(t). \tag{3.11}$$

In this task, we are looking at a very specific period of time, during which the gravitational experiment was performed. As mentioned in the introduction, this corresponds roughly to the month of June 2002, that is, 4 years and 9 months after launch. Given this short time frame, Eq. (3.10) shows a decrease of only 0.34 %, so that we can reasonably take the power as constant. Inserting $t = 4.75$ years into Eqs. (3.10) and (3.11), we obtain the reference values for the available power

$$\begin{aligned} W_{\text{Total}} &= 12521 \text{ W}, \\ W_{\text{elec}} &= 724 \text{ W}, \\ W_{\text{RTG}} &= 11797 \text{ W}. \end{aligned} \tag{3.12}$$

3.3 Results and Discussion

3.3.1 Baseline Scenarios

In order to get some insight on the influence of the different parameters, prior to a more thorough statistical analysis, we set out a number of scenarios that can be used as reference. In this section, a more accurate value of $m_{\text{Cassini}} = 4591$ kg is used for the spacecraft mass [10].

The simplest possible scenario, keeping in mind that the RTG contribution is expected to be the dominant one, is to simply consider their effect without reflection on the covers. This means that all power is absorbed and reemitted with the structure at a constant temperature. This results in an acceleration along the z-axis,

$$\mathbf{a}_{\text{th,Sc1}} = (-5.09\mathbf{e}_x - 8.81\mathbf{e}_y + 200\mathbf{e}_z) \times 10^{-11} \text{ m/s}^2. \tag{3.13}$$

This possibility represents the lower bound for this problem.

The next logical step is to include a small amount of reflection from the inner surface of the RTG shades. These structures are covered with a black Kaplan multilayer insulation (MLI), which has a high absorbance of around 90 % and also a high emittance of around 0.8. In terms of the Phong reflection formulation, this translates into a high diffusive reflection coefficient, of around 0.72 and a specular reflection coefficient of around 0.1. Tests conducted on the MLI during the development stages of the mission also show that the temperature on the inner layers remains low [11]. This also means that there is a small amount of power being transferred to the RTG cover's inner structure, precluding any significant power transfer to the main body through heat conduction from the RTG shades.

Translating this to our model, we first consider as a conservative estimate, a diffusive reflection coefficient of 0.4 and a specular reflection coefficient of 0.1. These conditions yield an acceleration of

$$\mathbf{a}_{\text{th,Sc2}} = (-6.87\mathbf{e}_x - 11.9\mathbf{e}_y + 269\mathbf{e}_z) \times 10^{-11} \text{ m/s}^2. \tag{3.14}$$

To obtain what would be an upper limit for the RTG contribution, we set the reflectivity coefficients at double the previous scenario, which would mean a total reflection of the thermal power irradiating the inner surface of the RTG covers. This hypothesis yields an acceleration of

$$\mathbf{a}_{\text{th,Sc3}} = (-8.65\mathbf{e}_x - 15.0\mathbf{e}_y + 337\mathbf{e}_z) \times 10^{-11}\ \text{m/s}^2. \quad (3.15)$$

If we add to the previous conditions, the upper bound for the contribution from the electrical equipment, meaning that all the power would be dissipated through the lower wall, we get a slightly larger acceleration on the z-axis,

$$\mathbf{a}_{\text{th,Sc4}} = (-8.65\mathbf{e}_x - 15.0\mathbf{e}_y + 372\mathbf{e}_z) \times 10^{-11}\ \text{m/s}^2. \quad (3.16)$$

This scenario gives us the upper limit for the overall acceleration given by our model.

A more reasonable scenario, is to take the second one considered above, using the reflection coefficients of 0.4 and 0.1, and add to it a contribution from the spacecraft body assuming that power is dissipated uniformly through all the surfaces. The MLI blanket covering the spacecraft body has the effect of evening out major temperature differences along the probe's structure, making this hypothesis reasonable. This scenario yields a small increase in the z component of the acceleration relative to Eq. (3.14),

$$\mathbf{a}_{\text{th,Sc5}} = (-6.87\mathbf{e}_x - 11.9\mathbf{e}_y + 272\mathbf{e}_z) \times 10^{-11}\ \text{m/s}^2. \quad (3.17)$$

This last set of hypotheses represents the baseline for the parametric study that follows in the next section. Notwithstanding, we can already point out that the z component is remarkably close to the value of $3 \times 10^{-9}\ \text{m/s}^2$, reported through the Doppler analysis [7].

The results from all the scenarios, which are described in Table 3.2, are summarised in Table 3.3.

In these results, the off-axis components remain about one order of magnitude below the values reported in Ref. [7]. However, the authors themselves point out that

Table 3.2 Summary of the scenarios used to obtain baseline results

Scenario	Description
1	Only RTG contribution, no refletion; lower bound
2	Only RTG contribution, low refletion
3	Only RTG contribution, high refletion
4	Scenario 3 with max. equipment contribution; upper bound
5	Scenario 2 with body contribution at uniform tamperature

Table 3.3 Summary of results from baseline scenarios for the Cassini thermal analysis

Scn.	W_{bottom} (W)	W_{top} (W)	$k_{\text{d,umbr}}$	$k_{\text{s,umbr}}$	a_x (10^{-11}m/s^2)	a_y (10^{-11}m/s^2)	a_z (10^{-11}m/s^2)
1	0	0	0	0	−5.09	−8.81	200
2	0	0	0.4	0.1	−6.87	−11.9	269
3	0	0	0.8	0.2	−8.65	−15.0	337
4	724	0	0.8	0.2	−8.65	−15.0	372
5	47.2	121	0.4	0.1	−6.87	−11.9	272

W_{bottom} and W_{top} are the emitted powers from the bottom wall of the lower module and the top wall of the main bus, respectively; $k_{\text{d,umbr}}$ and $k_{\text{s,umbr}}$ are the RTG umbrella reflection coefficients, and a_x, a_y and a_z are the components of the resulting thermal acceleration

the estimates of those components are not overly reliable. Furthermore, their values are presented relative to the orbital plane, whereas the results of this thermal analysis correspond to a reference frame aligned with the spacecraft.

One could speculate that this difference is due to the rotation between a reference with the z axis along the axis of the high-gain antenna and one with the z axis on the orbital plane. A simple calculation, hypothesising that the antenna is pointing directly towards Earth can be performed using data from the *Cassini, Galileo, and Voyager ephemeris tool* [12]. During the solar conjunction experiment, the angle between the two reference frames remained between 1.6° and 1.8°. The projection of the z component of the acceleration on the spacecraft frame on a direction orthogonal to the orbital plane would result in an acceleration component close to 10^{-10} m/s^2, which would agree with the order of magnitude of the Doppler measurements.

In the absence of more complete information on the methods used to obtain the Doppler estimates and the spacecraft orientation during the time of the experiment, it is not possible to make any definite assertions about the off-axis components of the acceleration.

3.3.2 Parametric Analysis

We now proceed to the statistical analysis based on a Monte Carlo simulation that aims to account for the uncertainties involved in the model. This analysis is performed only for the z component of the acceleration, since there is not enough information to properly constrain the relevant parameters for the off-axis components and, as discussed in the previous section, the results would not be reliable enough to draw any conclusions.

In the Monte Carlo method, a large number of random values associated with a statistical distribution are generated for each of the relevant parameters that influence the final result. This type of analysis is similar to the one performed in Chap. 2.

Table 3.4 Assumptions used to generate input parameter values for Monte Carlo simulation of Cassini thermal acceleration

Parameter	Symbol	Distribution
Diffusive reflec. coeff. of umbrella	$k_{\mathrm{d,umbr}}$	Normal, with $\mu = 0.6$ and $\sigma = 0.1$
Specular reflec. coeff. of umbrella	$k_{\mathrm{s,umbr}}$	Uniform, with interval $[0, 0.2]$
Diffusive reflec. coeff. of antenna back	$k_{\mathrm{d,ant}}$	Uniform, with interval $[0, 0.6]$
Specular reflec. coeff. of antenna back	$k_{\mathrm{s,ant}}$	Uniform, with interval $[0, 0.2]$
Power emitted from top of main bus	W_{top}	Uniform, with interval $[0, 94.4]$ W
Power from bottom of lower module	W_{bottom}	Uniform, with interval $[0, 242]$ W

The reflection coefficients of the RTG umbrellas can be reasonably constrained, as discussed in the preceding section, since we have some reliable data on the material and its properties [11]. This enables us to use a Normal distribution for the diffusive reflection coefficient, centred at a conservative estimate of 0.6 and with a σ of 0.1, allowing for a variation between 0.4 and 0.8 within 2σ. For all other parameters, we use uniform distributions with a reasonably wide interval in order to account for the lack of accurate and reliable data. The assumptions used to generate the random values for the simulation are outlined in Table 3.4.

Running a simulation with 10^5 iterations, we obtain the probability density distribution depicted in Fig. 3.6. The distribution is approximately normal. The mean of the resulting distribution is 3.01×10^{-9} m/s^2, with a standard deviation of 1.63×10^{-10} m/s^2. The acceleration along the Earth-spacecraft axis, with an uncertainty interval of 2σ, is

$$(a_{\mathrm{Cassini}})_z = (3.01 \pm 0.33) \times 10^{-9}\ \mathrm{m/s^2}, \tag{3.18}$$

which should be compared with 3×10^{-9} m/s^2 obtained by Bertotti et al. as an estimate of non gravitational acceleration from radiometric analysis [7].

From this analysis, one can conclude that the value for the thermal acceleration given by this model of the Cassini spacecraft is in agreement with the value obtained from the Doppler data, up to a 95 % probability level. It should be highlighted how close to the radiometric estimate the mean value of the thermal acceleration obtained here is.

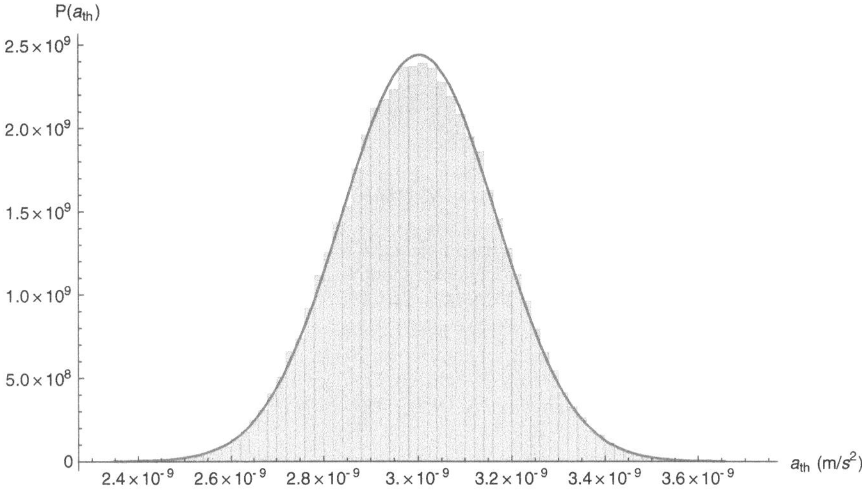

Fig. 3.6 Probability density distribution resulting from the Monte Carlo simulation of the thermal acceleration along the z-axis, with the normal distribution with the same mean and standard deviation superimposed

3.4 Conclusions

The results found in the modelling of the thermally induced acceleration of the Cassini space probe during its solar conjunction experiment significantly reinforced our confidence in the method first developed to account for the Pioneer anomaly (*cf.* Chap. 2). The adaptability of this approach allowed its application to an entirely new problem, with a different geometry, material properties and set of hypotheses, upholding the transparency and the simplicity of the method.

Clearly, some open questions still remain. The off-axis components of the acceleration are still poorly known. More detailed information about the internal power consumption and the attitude of the probe would be needed to properly address this issue. However, the result for the main component of the acceleration, along the probe's z axis, gives a very compelling result that closely agrees with the estimates of the non-gravitational acceleration presented in Ref. [7].

It becomes clear that any residual non-gravitational acceleration is negligible, since thermal effects account for practically the totality what was measured. This result significantly boost the confidence in one of the most accurate experiments ever performed to test General Relativity.

References

1. JPL Photojournal (2014), http://photojournal.jpl.nasa.gov/catalog/PIA03883. Accessed 02 Nov 2014
2. Cassini Mission to Saturn (2014), http://www.nasa.gov/mission_pages/cassini/main/index.html. Accessed 02 Nov 2014
3. C.M. Will, The confrontation between general relativity and experiment. Living Rev. Relativ. **9** (2006). doi:10.12942/lrr-2006-3
4. O. Bertolami, J.Páramos, S.G. Turyshev, in *General Theory of Relativity: Will It Survive the Next Decade?*. Lasers, Clocks and Drag-Free Control (Springer, Berlin, 2008), pp. 27–74. ISBN 978-3-540-34376-9. doi:10.1007/978-3-540-34377-6-2
5. O. Bertolami, J.Páramos, in *The Experimental Status of Special and General Relativity*. Springer Handbook of Spacetime (Springer, Berlin, 2014), pp. 463–483
6. L. Iess, G. Giampieri, J.D. Anderson, B. Bertotti, Doppler measurement of the solar gravitational deflection. Class. Quantum Gravity **16**(5), 1487–1502 (1999). doi:10.1088/0264-9381/16/5/303
7. B. Bertotti, L. Iess, P. Tortora, A test of general relativity using radio links with the Cassini spacecraft. Nature **425**(6956), 374–376 (2003). doi:10.1038/nature01997
8. O. Bertolami, F. Francisco, P.J.S. Gil, J. Páramos, Modeling the nongravitational acceleration during Cassini's gravitation experiments. Phys. Rev. D. **90**(4), 042004 (2014). doi:10.1103/PhysRevD.90.042004
9. P.S. Cooper, Searching for modifications to the exponential radioactive decay law with the Cassini spacecraft. Astropart. Phys. **31**(4), 267–269 (2009). doi:10.1016/j.astropartphys.2009.02.005
10. M. Di Benedetto, The non-gravitational accelerations of the Cassini spacecraft and the nature of the Pioneer anomaly, Ph.D. thesis, Spienza - Universitá di Roma, October 2011
11. E.I. Lin, J.W. Stultz, Cassini multilayer insulation blanket high-temperature exposure tests. J. Thermophys. Heat Transf. **9**(4), 778–783 (1995). doi:10.2514/3.738
12. Cassini, Galileo, and Voyager ephemeris tool (2014), http://www-pw.physics.uiowa.edu/jbg/cas.html. Accessed 02 Nov 2014

Chapter 4
Outer Solar System (OSS) Mission Proposal

4.1 Overview

The connection between gravitational physics and solar system physics is, certainly, one of the underlying themes of this thesis. Especially, in the issues discussed in the previous two chapters, this connection is prominently in display, as both concern space missions that had planetary science at the heart of their scientific objectives, but created some of the best opportunities so far to study gravity in the solar system. It is also somewhat surprising that, in both cases, thermal engineering analysis turned out to have such an importance to the full understanding of the gravitational experiment results.

The proposed Outer Solar System (OSS) mission [1] continues this tradition by associating the communities of fundamental physics and planetary sciences in a mission with ambitious goals in both fields. OSS would visit Neptune and its moon Triton, nearly half a century after Voyager 2 did it for the first and only time. Using a suite of advanced instrumentation with a strong heritage from previous outer solar system missions, OSS would provide striking advances in the study of the farthest known planet of the solar system. The Neptune flyby would be precisely controlled to permit a close encounter with a Kuiper Belt Object (KBO) to be properly chosen among the large number of scientifically interesting and attainable objects (owing to the large mass of Neptune, this number being much larger than for New Horizons, NASAs fast-track mission towards Pluto and a KBO). A mission like OSS would have the potential to consolidate the hypothesis of the origin of Triton as a KBO captured by Neptune at the time of formation of the solar system.

The OSS probe would carry instruments allowing for a precise tracking of the spacecraft during the cruise stage of the mission. It would make possible the best tests ever of the laws of gravity in the outer solar system, more than a hundred times more accurately than currently done. This is important, not only for fundamental physics, but also for cosmology and astrophysics in a context where the observations currently interpreted in terms of dark matter and dark energy, challenge General Relativity (GR) at scales much larger than that of the solar system. The scientific

F. Francisco, *Trajectory Anomalies in Interplanetary Spacecraft*, Springer Theses,
DOI 10.1007/978-3-319-18980-2_4

goal of better tests of the laws of gravity is also directly connected to the question of the origin of the solar system, as models of its formation use GR to describe the crucial role of gravity. Using laser metrology, the OSS mission will also improve the result of the Cassini spacecraft that measured the γ parameter during its interplanetary journey to Saturn (*cf.* Chap. 3).

In this chapter, we briefly address the main features of this mission proposal, for which the author of this thesis was a contributor, with a natural highlight to the scientific objectives related with fundamental physics. Full details on the proposal can be found in Ref. [1].

4.2 Scientific Objectives

4.2.1 Deep Space Gravity

General Relativity, the current theoretical formulation of gravitation, is in good agreement with experimental tests of gravitation so far [2–4]. Meanwhile, the experimental tests leave open windows for deviations from General Relativity at short or long distance scales [5, 6].

Testing gravity at the largest scales reachable by man-made instruments is therefore essential to bridge the gap between experiments in the solar system and astrophysical or cosmological observations. The most notable existing test in this domain was performed by NASA during the extended Pioneer 10 and 11 missions. This test resulted in what is now known as the Pioneer anomaly, one of the few experimental signals apparently deviating from the predictions of General Relativity (*cf.* Chap. 2).

Several mission concepts have been put forward to improve the experiment performed by Pioneer 10 and 11 probes. A key idea in these proposals is to measure non-gravitational forces acting on the spacecraft, independently of their underlying cause. This requires a very tight control of all systematic effects [7, 8].

The addition of an accelerometer on board the spacecraft, not only improves the precision and quality of the navigation, but also allows for a better understanding of the origin of any anomalous signals that might surface. The target accuracy of the acceleration measurement is $10^{-12}\ \mathrm{m/s^2}$ after an integration time of 3 h. Combining these measurements with radio tracking data, it becomes possible to improve by three orders of magnitude the precision of the comparison with theory of the spacecraft gravitational acceleration, when compared with the data obtained from the Pioneer 10 and 11 tracking.

The same instrument also improves the science return with respect to objectives in exploration of the outer solar system physics, which is the motivation for combining fundamental physics and planetary physics in a common mission. This idea has been included in the Roadmap for Fundamental Physics in Space issued in 2010 by ESA [9].

4.2.2 Measurement of the γ Parameter

Metric extensions of General Relativity are often characterised in terms of the Parameterised Post Newtonian (PPN) formalism [2]. In particular, the γ parameter can be used to gauge the fractional strength of scalar interaction in scalar-tensor theories of gravity. This deviation of this parameter from unity has been shown to be smaller than 2×10^{-5} by the Cassini relativity experiment performed at solar conjunctions in June 2002 (*cf.* Chap. 3). But recent theoretical proposals suggest that this deviation might have a natural value in the range of 10^{-6} to 10^{-7} as a consequence of a damping of the scalar contribution to gravity during cosmological evolution [10].

The purpose is to use the same setup as in the Cassini solar conjunction experiments, with the advantage of an accelerometer on board to measure the non-geodesic acceleration. A largely improved accuracy can be attained with the up-scaling option of a laser ranging equipment onboard. The OSS mission can thus measure the γ parameter at the 10^{-7} level, which would provide new crucial information on scalar-tensor theories of gravity at their fascinating interface with theories of cosmological evolution.

4.2.3 Other Objectives

The OSS mission includes a wide range of other scientific objectives, mainly related with planetary science, that are briefly outlined here. A more detailed explanation of these objectives can be found in Ref. [1].

4.2.3.1 Neptune's Interior

The interior of Neptune is poorly understood, but likely composed of a mixture of rock and ices. It is not clear, however, if rock and ice components are fully or incompletely separated. The radial extent of the core region could amount up to 70 % of the total radius, thereby substantially affecting the planets gravitational field. A better knowledge of Neptune's shape and rotational state would allow for a better constraint of the models of its interior.

4.2.3.2 Neptune's Atmosphere

Despite its great distance from the sun, Neptune has a surprisingly dynamic atmosphere, including a jet stream blowing at almost 500 m/s and a giant vortex [11]. The great question is how this dynamical weather system is powered.

High-resolution observations with a camera optimised for Neptunes atmosphere will enable a search for eddies at the relevant spatial scales. Similarly measurements of Neptunes thermal emission from the mid-infrared through the sub-millimetre can determine the depth to which differences between both axisymmetric and discrete regions exist. These are essential measurements to determine what powers Neptune's circulation and thermal structure.

4.2.3.3 Neptune's Magnetic Field

Neptune's magnetic dipole, like that of Uranus, is highly tilted and offset from the planet's centre. Neptune's magnetic field also goes through dramatic changes as the planet rotates in the solar wind, with the magnetosphere being completely reconfigured twice per planetary rotation period [12]. Thus, it is not clear why, despite this, the magnetosphere appeared very quiescent during the Voyager 2 flyby in 1989.

4.2.3.4 Triton

Another major science goal is to unravel Triton's geological history based on imaging science observations with high spatial resolution. This has been prevented to date because of the limited coverage and low spatial resolution of most images collected during the Voyager flyby. The few higher resolution images reveal a geologically young, complex surface unlike any other seen in the outer solar system.

4.2.3.5 Neptune's Rings and Inner Satellites

The OSS science payload and flyby trajectory offer a unique opportunity to increase our understanding of the Neptunian ring system and its group of small inner satellites. Ring-moon systems were once perceived as stable and unchanging for time scales of at least 10^6 years. New, higher-quality data from Cassini, Hubble and ground-based telescopes are painting a different picture, in which the systems evolve over years to decades [13]. A good coverage in the OSS flyby will be the key to infer the size distribution from visual and near-infrared observations of the rings. The dust detector can directly determine the composition of micron sized grains in the extended Neptunian dust disk from in situ measurements.

4.2.3.6 Kuiper Belt Objects

Our understanding of the history of our solar system has been revolutionised, largely because of the discovery of Kuiper Belt Objects (KBOs). Well over a thousand KBOs have since been discovered and they exhibit remarkable diversity in their properties [14]. The overall objective is to guarantee that the necessary data for detailed comparison of Triton and the OSS KBO to each other, and to Pluto and the

New Horizons KBO is obtained. Together, these observations will begin to provide significant insight into the diverse KBO population.

4.3 Mission Profile

The capability to embark the necessary instrument suite is highly dependent on the chosen orbit and launcher. A preliminary analysis of a mission profile capable of delivering a 500 kg class probe is here outlined.

The selected orbit shall meet the following criteria:

(i) Long ballistic periods to follow geodesic arcs to as large a heliocentric distance as possible for gravity measurements;
(ii) Sun occultation for γ parameter measurement;
(iii) Neptune and Triton flybys for planetary science measurements, then a flyby of a scientifically selected Kuiper Belt object;
(iv) Transfer to Neptune in less than 13 years for a rapid science return;
(v) Mission Δv, and thus the onboard propellant quantity, as small as possible to increase the delivered mass;
(vi) Low departure velocity to reduce launch cost.

The first requirement is naturally achieved by the Neptune objective, leading to heliocentric distance larger than 30 AU, with possible extension after Neptune flyby.

For the second requirement, since Neptune's orbit is not in the ecliptic plane, the Earth-Sun-spacecraft conjunctions can occur only early in the mission or when the spacecraft is near the ecliptic plane. With direct transfers, two conjunctions will occur, at 2.1 AU (6 months after departure) and at 4.3 AU. An indirect transfer increases the number of solar conjunction during inner solar system trajectory.

The Neptune encounter provides access to a huge cone of trans-Neptunian space in order to achieve the third requirement. The bending angle that can be achieved varies with the approach velocity to Neptune and the trajectorys minimum distance from Neptune. A faster and higher trajectory decreases the angle while a slower one could increase it. Trajectory modelling shows that tens of known KBOs are accessible to OSS, and the flyby geometry can be tailored to not only achieve science goals within the Neptune system, but continue on to a selected KBO afterward.

For the last three requirements, different strategies have been analysed. A direct trajectory would enable almost annual launch windows at the expense of a relatively heavy launcher due to the high initial velocity required. Transfers using inner solar system gravity assists would allow less heavy launchers. Two optimised trajectories are compared in Fig. 4.1.

It is expected that during Neptune and Triton encounter, the flyby velocity is less than the one for New Horizons for Pluto-Charon encounter, with the same type of instrumentation and scientific objectives. The time allocation between the difference experiments should be analysed during further steps.

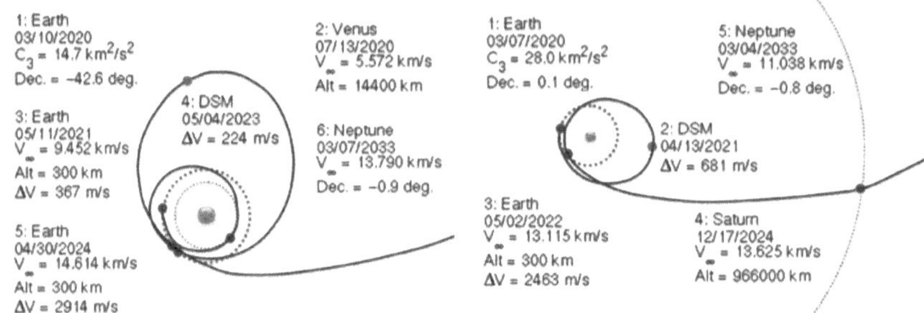

Fig. 4.1 OSS orbit for a launch in 2020, with Venus and 2 Earth gravity assists (VEEGA) on the *left* and 2 Earth and Saturn gravity assists (EESGA) on the *right*

4.4 Conclusions

The main fundamental physics objective is the deep space gravity test driven by the research of deviations from GR. In a context dominated by the quest for the nature of dark matter and energy, testing gravity at the largest scales reachable by man-made instruments is essential to bridge the gap with astrophysical and cosmological observations. For that purpose, it is necessary to have the best possible navigational accuracy during the interplanetary trajectory. this can be achieved by combining a high precision accelerometer with the conventional radiometric techniques. The fundamental physics tests also allow for a scientific return long before the arrival at Neptune, which occurs 13 years after launch, without competing with the planetary instruments for power or data bandwidth.

The constraints on mass and cost of the mission imposed by ESA lead to the choice of a flyby of Neptune and Triton instead of a Triton orbital insertion. Nevertheless, this choice allows for the continuation of the mission with a flyby of a KBO, enabling for the discrimination of hypothesis regarding the origin of Triton as an object captured by Neptune at the time of formation of the solar system.

The interest of visiting Neptune again and Triton after the Voyager 2 flyby, half a century ago, relies on ground-based observations of time-variable phenomena in the planets atmosphere and the ring system. The scientific return of a mission like OSS could fuel our knowledge of the farthest planet of the solar system owing to substantial technological progress made since the days of Voyager 2. Furthermore, the mission would very much benefit from the maturity of existing instruments, as the OSS planetary objectives can be addressed by an instrument suite similar to that flown onboard the New Horizons spacecraft that is presently underway to Pluto and its satellites.

Despite its scientific interest and its technological appeal, the OSS mission barely missed ESA's selection process. It is clear, however, that the main objectives of the OSS mission remain quite relevant and can be of interest in future calls for innovative and scientifically well motivated deep space missions.

References

1. B. Christophe, L.J. Spilker, J. Anderson, N. André, S.W. Asmar, J. Aurnou, D. Banfield, A. Barucci, O. Bertolami, R. Bingham, P. Brown, B. Cecconi, J.M. Courty, H. Dittus, L.N. Fletcher, B. Foulon, F. Francisco, P.J.S. Gil, K.H. Glassmeier, W. Grundy, C. Hansen, J. Helbert, R. Helled, H. Hussmann, B. Lamine, C. Lämmerzahl, L. Lamy, R. Lehoucq, B. Lenoir, A. Levy, G. Orton, J. Páramos, J. Poncy, F. Postberg, S.V. Progrebenko, K.R. Reh, S. Reynaud, C. Robert, E. Samain, J. Saur, K.M. Sayanagi, N. Schmitz, H. Selig, F. Sohl, T.R. Spilker, R. Srama, K. Stephan, P. Touboul, and P. Wolf, OSS (Outer Solar System): a fundamental and planetary physics mission to Neptune, Triton and the Kuiper Belt. Exp. Astron. **34**(2), 203–242 (2012). doi:10.1007/s10686-012-9309-y
2. C.M. Will, The confrontation between general relativity and experiment. Living Rev. Relativ. **9** (2006). doi:10.12942/lrr-2006-3
3. O. Bertolami, J. Páramos, S.G. Turyshev, *General Theory of Relativity: Will It Survive the Next Decade?*. Lasers, Clocks and Drag-Free Control (Springer, Berlin, 2008), pp. 27–74. ISBN 978-3-540-34376-9. doi:10.1007/978-3-540-34377-6-2
4. O. Bertolami, J. Páramos, *The Experimental Status of Special and General Relativity*. Springer Handbook of Spacetime (Springer, Berlin, 2014), pp. 463–483
5. E.G. Adelberger, B.R. Heckel, A.E. Nelson, Tests of the gravitational inverse-square law. Annu. Rev. Nucl. Part. Sci. **53**(1), 77–121 (2003). doi:10.1146/annurev.nucl.53.041002.110503
6. S. Reynaud, M.-T. Jaekel, Testing the Newton law at long distances. Int. J. Mod. Phys. A **20**(11), 2294–2303 (2005). doi:10.1142/S0217751X05024523
7. O. Bertalmai, J. Páramos, A mission to test the Pioneer anomaly: Estimating the main systematic effects. Int. J. Mod. Phys. D **16**, 1611 (2007). doi:10.1142/S0218271807011000
8. J.D. Anderson, S.G. Turyshev, M. Nieto, A mission to test the Pioneer anomaly. Int. J. Mod. Phys. D **11**(10), 1545 (2002). doi:10.1142/S0218271802002876
9. A Roadmap for Fundamental Physics in Space (2014), http://sci.esa.int/fprat. Accessed 18 Nov 2014
10. T. Damour, F. Piazza, G. Veneziano, Runaway dilaton and equivalence principle violations. Phys. Rev. Lett. **89**(8), 081601 (2002). doi:10.1103/PhysRevLett.89.081601
11. B.A. Smith, L.A. Soderblom, D. Banfield, C. Barnet, R.F. Beebe, A.T. Bazilevskii, K. Bollinger, J.M. Boyce, G.A. Briggs, A. Brahic, Voyager 2 at Neptune-imaging science results. Science **246**, 1422–1449 (1989). doi:10.1126/science.246.4936.1422
12. F. Bagenal, Giant planet magnetospheres. Annu. Rev. Earth Planet. Sci. **20**, 289–328 (1992). doi:10.1146/annurev.ea.20.050192.001445
13. C.D. Murray, K. Beurle, N.J. Cooper, M.W. Evans, G.A. Williams, S. Charnoz, The determination of the structure of Saturn's F ring by nearby moonlets. Nature **453**, 739–744 (2008)
14. M.A. Barucci, H. Boehnhardt, D.P. Cruikshank, A. Morbidelli, The Solar System Beyond Neptune: Overview and Perspectives. The Solar System Beyond Neptune, pp. 3–10. University of Arizona Press, 2008

Chapter 5
The Flyby Anomaly and Options for Its Study

5.1 Background

In the past couple of decades, a few deep-space probes that used Earth gravity assists have apparently displayed an unexpected change in their hyperbolic excess velocities. This issue had already been the subject of discussion since the mid 1990s [1], but had maintained a relatively low profile in the scholarly journals until a paper came out by Anderson et al. [2]. This was at the height of the Pioneer anomaly controversy, so the physics community was especially receptive to the discussion of spacecraft trajectory anomalies, However, unlike the Pioneer anomaly, this so-called *flyby anomaly* remains an open question ever since.

A gravitational assist or *flyby* is a manoeuvre often used in interplanetary missions to save fuel. It involves a relatively low altitude and high speed flyby around a solar system body, usually a planet. The spacecraft follows a hyperbolic trajectory around the planet, like the one illustrated in Fig. 5.1, changing its direction and speed and, consequently, its heliocentric orbital parameters. Earth has often been used to perform these manoeuvres, notably, in the missions where the flyby anomaly was allegedly detected.

This anomaly was detected in the residuals of the radiometric tracking data of several space probes performing Earth flybys. The trajectories inferred from the tracking data proved impossible to fit to a single hyperbolic arc. The pre-encounter and post-encounter data had to fit to separate hyperbolic trajectories that displayed a discrepancy in the hyperbolic excess velocities. Since the tho hyperbolic arcs appear to be good fits for their respective data sets, it is assumed that this velocity shift is localised near the perigee, where tracking through the Deep Space Network (DSN) is not available for approximately 4 h [2]. The spatial resolution of the available reconstructions, resulting from the 10 s interval tracking, does not allow for an accurate characterisation of the effect, so that no corresponding spatial or temporal profile of the acceleration exists. The only available data to support the analysis is the shift in excess velocity (and correspondingly, in kinetic energy) between the inbound and outbound trajectories.

F. Francisco, *Trajectory Anomalies in Interplanetary Spacecraft*, Springer Theses,
DOI 10.1007/978-3-319-18980-2_5

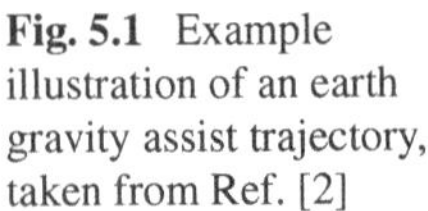

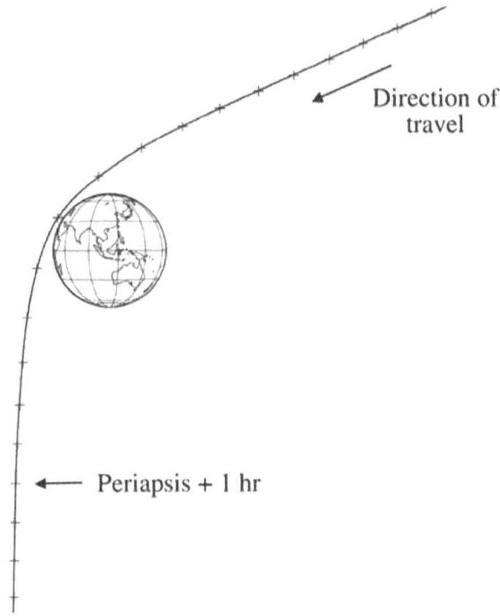

Fig. 5.1 Example illustration of an earth gravity assist trajectory, taken from Ref. [2]

The flyby anomaly has been observed in the Galileo, NEAR, Rosetta and Cassini missions. Earth flybys between 1990 and 2005 are listed in Table 5.1, based on data from Ref. [2], with respective perigee altitudes (h_p) and velocities (v_p), hyperbolic excess velocities (v_∞) and the anomalous shift in excess velocity (Δv_∞).

It should be noted that the information on the flybys where the anomaly was allegedly detected is not entirely consistent across the different sources. Aside from minor disparities in the values of some parameters, there are a few important differences that deserve some discussion.

Regarding to the second Galileo flyby, Ref. [2] explains that the measured value for the change in excess velocity (Δv_∞) was actually −8 and the −4.6 mm/s figure in Table 5.1 is obtained after subtracting an estimate of the atmospheric drag. However, for the same flyby, two other sources claim that, due to the low altitude, atmospheric drag would mask any anomalous velocity change [1, 3]. The reasons for absence of any value for the August 1999 Cassini flyby in Ref. [3] are also uncertain, as no explanation is offered for this omission.

Table 5.1 List of Earth flybys between 1990 and 2005, based on data from Ref. [2]

Date	Mission	h_p (km)	v_p (km/s)	v_∞ (km/s)	Δv_∞ (mm/s)
08/12/1990	Galileo	960	13.740	8.949	3.92
08/12/1992	Galileo	303	14.080	8.877	−4.6
23/01/1998	NEAR	539	12.739	6.851	13.46
18/08/1999	Cassini	1175	19.026	16.01	−2
04/03/2005	Rosetta	1956	10.517	3.863	1.80
02/08/2005	MESSENGER	2347	10.389	4.056	0.02

h_p and v_p are the altitude and velocity at the perigee, v_∞ is the hyperbolic excess velocity and Δv_∞ is the measured change in excess velocity

Table 5.2 Recent earth flybys for which no data is yet available in the literature

Date	Mission
13/11/2007	Rosetta
13/11/2009	Rosetta
09/10/2013	Juno

When one compares the tables presented in Refs. [2, 3], one is not merely presented with slight differences in values. The two sets of data in these two papers could actually present a whole different picture on the phenomenon, potentially prompting different assumptions. While in the first the effect appears to be bidirectional, accelerating the spacecraft on some occasions and decelerating in others, in the second reference all instances that have negative velocity changes have been removed and the effect appears to be consistent with an energy increase.

The study of more flybys and closure of the 4 h gap in DSN tracking near the perigee would be essential to shed some more light onto this phenomenon. There have been a few recent flybys, listed in Table 5.2, however their results are yet to be presented in scholarly literature.

At this point, it must be acknowledged that the flyby anomaly is still very poorly characterised and any consistent treatment is, at least, very difficult. These limitations clearly raise the need for alternative approaches.

The discussion that follows in this chapter aims to formulate a proposal of a new experimental method to study the flyby anomaly. Specifically, we explore the possibility of using Global Navigation Satellite System (GNSS) constellations to study the flyby anomaly. GNSS satellite tracking would allow, not only to obtain orbital solutions, but also to make a possible distinction between anomalies in the actual trajectories and anomalies in the propagation of light [4].

In the remainder of the chapter we shall build the concept of a mission to study the flyby anomaly [5]. To do this, we begin by discussing some of the theories put forward in attempts to explain the anomaly as well as the possible effects of each of those possibilities in GNSS systems. Afterwards, the technical possibilities of GNSS spacecraft tracking are addressed and, finally, the mission requirements are outlined.

5.2 Proposed Explanations

5.2.1 Systematic Effects

When presented with any unexpected effect, the first step is always to search for systematic effects that could possibly explain it.

The detailed discussion of the two Galileo (1990 and 1992) and the NEAR (1998) gravity assists made in Ref. [1] includes an account of the accelerations generated by different known effects, in an attempt to single out possible error sources. In that paper, an estimated average acceleration associated with the flyby anomaly of the order of 10^{-4} m/s^2 is measured against the Earth oblateness, other Solar System bodies, relativistic corrections, atmospheric drag, Earth albedo and infrared emissions, ocean tides, solar pressure, etc. From this analysis, the authors conclude that the hypothetical acceleration would have a magnitude second only to the Earth oblateness.

Subsequently, this discussion was extended to other possible error sources, comparing this 10^{-4} m/s^2 figure with several additional unaccounted acceleration sources. These include the atmosphere, ocean tides, solid tides, spacecraft charging, magnetic moments, Earth albedo, solar wind and spin-rotation coupling. It is concluded that all of the considered effects are several orders of magnitude smaller than the flyby anomaly [6].

A study of the thermal effects in the first Rosetta flyby has also been preformed, concluding that they cannot be responsible for the reported flyby anomaly [7]. This discussion is somewhat reminiscent of the one surrounding the Pioneer anomaly, for which the thermal effects argument, not only presents a much more solid case, but ultimately turned out to be the key for its understanting [8, 9], as discussed in Chap. 2 of this thesis.

A quick overview of the magnitudes of all effects discussed in the preceding paragraphs is compiled in Table 5.3. It is clear that all listed effects except the Earth oblateness are orders of magnitude smaller than the required value. This raises the issue of possible errors in the gravitational model of the Earth. However, attempts to solve the flyby problem by changing the related second dynamic form factor J_2 have yielded unreasonable solutions, and are unable to account for all flybys [1].

Table 5.3 List of possible error sources and respective orders of magnitude during earth flybys

Effect	Order of magnitude (m/s^2)
Earth oblateness	10^{-2}
Other solar system bodies	10^{-5}
Relativistic effects	10^{-7}
Atmospheric drag	10^{-7}
Ocean and earth tides	10^{-7}
Solar pressure	10^{-7}
Earth infrared	10^{-7}
Spacecraft charge	10^{-8}
Earth albedo	10^{-9}
Solar wind	10^{-9}
Magnetic moment	10^{-15}

Values should be compared with an estimated anomalous acceleration of 10^{-4} m/s^2

5.2.2 Earth Rotation

An empirical formula has been proposed by Anderson et al. in Ref. [2] to fit the flyby relative velocity change, $(\Delta v_\infty / v_\infty)$, as a function of the declinations of the incoming and outgoing asymptotic velocity vectors, δ_i and δ_o, respectively

$$\frac{\Delta v_\infty}{v_\infty} = K(\cos\delta_i - \cos\delta_o), \tag{5.1}$$

where the constant K is expressed in terms of the Earth's rotation velocity, $\omega_\oplus$; its radius, $R_\oplus$, and the speed of light, c, as

$$K = \frac{2\omega_\oplus R_\oplus}{c}. \tag{5.2}$$

This identification is suggestive, as it evokes the general form of the outer metric due to a rotating body,

$$ds^2 = \left(1 + 2\frac{V(r) - \Phi_0}{c^2}\right)(c\,dt)^2 - \left(1 - 2\frac{V(r)}{c^2}\right)(dr^2 + r^2 d\Omega^2), \tag{5.3}$$

with

$$\frac{\Phi_0}{c^2} = \frac{V_0}{c^2} - \frac{1}{2}\left(\frac{\omega_\oplus R_\oplus}{c}\right)^2, \tag{5.4}$$

where $d\Omega^2 = d\theta^2 + \sin^2\theta d\phi^2$, V_0 is the value of the Newtonian potential, $V(r)$, at the equator [10].

Following this reasoning, and given the strong latitude dependence of Eq. (5.1), this expression appears to indicate that the Earth's rotation might be generating a much larger effect than the frame dragging predicted by General Relativity (GR). This, however, is in contradiction with the recent measurements of this effect performed by Gravity Probe B, which orbits the Earth at a height of $\sim$600 km, well within the onset zone of the reported flyby anomaly [11].

5.2.3 Interaction with Dark Matter

Another putative explanation for the flyby anomaly relies on an additional drag experienced by spacecraft due to the presence of dark matter in Earth's vicinity [12]. Indeed, while drag between normal matter is proportional to the spacecraft's velocity and area, dark matter "drag" is proportional to the mass of the latter and may be negative, causing an increase in velocity.

The requirement that some flyby anomalies exhibit a positive Δv_∞ while others experience a negative shift (see Table 5.1) places strong constraints on the density profile of this dark matter component, namely that the latter should be endowed with inelastic forward dominated scattering, and of either elastic or inelastic backward dominated scattering. This could warrant a two-component dark matter model with an approximately isotropic velocity distribution, or a single dark matter component with an anisotropic velocity distribution, subject to inelastic as well as elastic scattering.

Furthermore, the scale of the typical acceleration of the flyby anomalies requires that the dark matter density is orders of magnitude above those considered for galactic dark matter halos, $\rho_{DM} \gg \rho_{halo} \sim 0.3$ (GeV/c^2)/cm^3. Indeed, assuming that dark matter particles have a mass well below 1 GeV/c^2 and a cross section of the order of $\sigma \sim 1$ mb, Adler obtains in Ref. [12] $\rho_{DM} \sim 10^{11}$ (GeV/c^2)/cm^3 and $\rho_{DM} \sim 10^7$ (GeV/c^2)/cm^3 for the elastic and inelastic scenarios, respectively. However, this hypothesis hangs on the existence of a suitable gravitational mechanism that allows for such a buildup of dark matter in the vicinity of the Earth.

Given the above discussion, Ref. [12] and the subsequent analysis by the same author in Refs. [13, 14] consider an array of constraints on either $\sigma\rho_{DM}$ or ρ_{DM} alone, arising from several astrophysical scenarios and attempt to derive a suitable dark matter density profile. These conclude that dark matter may be the cause of the flyby anomalies, although this would require severe constraints on its spatial distribution, something which could perhaps show up as low level signal modulations in current and future dark matter searches as DAMA/LIBRA [15].

In the present context, a confirmation and thorough spatial characterisation of the existence of the flyby anomaly would allow one to further constrain the density profile and physical characteristics of dark matter particles.

5.2.4 Modified Inertia

Another candidate to account for the flyby anomaly relies on a modification of inertia due to a form of Unruh radiation, with a Hubble-scale Casimir effect producing a difference between inertial m_I and gravitational mass m_G of the form

$$m_I = m_G \left(1 - \frac{\pi^2}{2}\frac{\beta c H}{a}\right), \tag{5.5}$$

where $\beta = 0.2$ is related to Wien's constant, H is Hubble's constant and a is the acceleration measured with respect to the local matter distribution [16]. This modification has already been explored in the search of a solution for the Pioneer anomaly [17].

This modification of inertial mass disagrees with tests of the equivalence principle (see e.g. Refs. [18, 19] for a review), leading its proponents to consider that, for some unknown reason, this phenomenological model only applies to unbound orbits.

With this caveat in mind, a straightforward analysis of the Earth-spacecraft energy exchange during perigee shows that an anomalous velocity shift arises,

$$\Delta v = \frac{\pi^2}{0.14} \frac{\beta R_\oplus c H}{v_\oplus^2} \frac{(v_2 \cos\phi_1 - v_1 \cos\phi_2)}{\cos\phi_1 \cos\phi_2}, \tag{5.6}$$

where $v_\oplus$ is the rotational velocity at the surface equator and v_i, ϕ_i are the incoming ($i = 1$) and outgoing ($i = 2$) velocity and latitudes. A somewhat lower quality of fit is obtained when compared with the similar looking Eq. (5.1), although it is argued that this is to be expected, as the latter is obtained from the fit itself, while Eq. (5.6) is derived from the underlying model here discussed.

Clearly, the availability of a high-resolution reconstruction of the flyby trajectories would allow for a much more detailed inspection of the consequences of the modified inertia model discussed above. Specifically, instead of simply considering the inbound and outbound velocities v_1 and v_2, one could directly solve the equation of motion

$$\frac{m_I}{m_G}\mathbf{a} = \left(1 - \frac{\pi^2}{2}\frac{\beta c H}{a}\right)\mathbf{a} = \frac{GM_\oplus}{r^2}\mathbf{e}_r, \tag{5.7}$$

and compare the computed trajectory with observations, thus validating or disproving the underlying model.

5.2.5 *Modified Particle Dynamics*

An axiomatic scheme developed to analyse how deviations from GR affect particle dynamics has been set up in Ref. [6], based upon a general *ansatz* for the metric and the related equation of motion that preserves the weak equivalence principle.

This proposal follows a similar spirit to the widely used Parameterised Post-Newtonian (PPN) formalism [20]. Regarding the latter, experimental data strongly constrain the PPN parameters to approach their GR values [18, 19], notably the measurements made by Cassini provide the current best constrain for the γ parameter [21, 22], as discussed in Chap. 3. These constrains would imply that any related effects are below 10^{-7} m/s^2, the order of magnitude of the relativistic ones.

However, this novel approach offers some additional contributions not considered in the PPN formulation, thus allowing for a wider range of phenomenological implications. These could include the Pioneer and flyby anomalies, amongst other phenomena, due to an anomalous acceleration of the form

$$\mathbf{a}_A = \frac{GM}{r^2}\left(A_{21}\frac{\mathbf{r}\cdot\mathbf{v}}{rc}\mathbf{e}_r + A_{22}\frac{\mathbf{v}}{c}\right), \tag{5.8}$$

plus higher order terms in v/c. Evaluating the above for a flyby trajectory, it is found that an additional along-track term dominates,

$$\mathbf{a}_A = A_{22}\frac{GM}{r^2}\frac{\mathbf{v}}{c}. \tag{5.9}$$

Indeed, it is shown that if the coefficient $A_{22} \sim 1$, the perigee altitudes and velocities reported in the affected spacecraft (see Table 5.1) yield an acceleration of the order of 10^{-4} m/s^2, typical of the flyby anomaly [6].

Again, the use of GNSS as an independent means of tracking gravity assists would be useful not only to confirm the existence of the anomaly, but also to test the modified trajectories derived from this additional along-track term, and perhaps of other subdominant terms. Similar effects in the propagation of light could also be tested through the statistical approach outlined before.

5.3 Effect on GNSS Systems

In order to discuss the possible use of current and future GNSS constellations to probe the flyby anomaly, one should first evaluate to what extent its hypothetical cause could affect the individual elements of the navigation system itself.

The typical navigation satellite follows an approximately circular Medium Earth Orbit (MEO), at a height of about 20,000 km. Since the anomalous velocity change is only observed before and after flybys occurring at much lower altitudes (of the order of 1000 km), one may empirically dismiss any effect.

The above argument can be sharpened, even though a full analysis is impossible due to the lack of spatial resolution and consequent inability to fully characterise the spatial dependence of the reported anomaly. Notwithstanding, one takes as relevant figure of merit the anomalous acceleration $a_{\text{flyby}} \sim 10^{-4}$ m/s^2, which may be assumed constant in the absence of a better description. In this case it has been shown that no constant acceleration greater than 10^{-9} m/s^2 can affect the GNSS constellation, since it would have otherwise been detected [23].

Thus, one concludes that the flyby anomaly, if caused by a distance dependent force, must be strongly decaying, dropping by four orders of magnitude with a modest (about fourfold) increase in distance, from $r = R_\oplus + h \simeq 7000$ km to $r \simeq 27{,}000$ km.

5.3.1 Interaction with Dark Matter

If the flyby anomaly is caused by dark matter "drag", as discussed in Sect. 5.2.3, the relevance of its effect on the GNSS follows an argument similar to the one above. One bound discussed in Ref. [12] is obtained from the stability of the LAGEOS geodetic satellite [24, 25], which orbits the Earth at an altitude of ~12,300 km and would experience a maximum drag acceleration of $10^{-12}\,\mathrm{m/s^2}$. Since the GNSS constellations are at approximately twice that height, and assuming a non-increasing density, one concludes that it should experience a drag below this value, as the dark matter drag is proportional to the mass of the spacecraft and each individual satellite has about the same mass as LAGEOS.

Furthermore, this is based on the hypothesis of the concentration of dark matter inside a shell with outer radius ~70,000 km. A recent refinement of this proposal drops this assumption, so that dark matter can be packed much closer to the Earth, well below the aforementioned height ~20,000 km of the GNSS. In either case, this possibility poses no concern to our analysis [14].

5.3.2 Modified Inertia

The other candidate solution for the flyby anomaly, addressed in Sect. 5.2.4, offers a different challenge. Indeed, this does not postulate a distance dependent force, but instead one that reflects the acceleration dependence of the inertial to gravitational mass ratio. As discussed before, it appears that this model, if viable, does not apply to planets, leading its proponents to judge it to be relevant for unbound orbits only.

One could follow suit and simply state that the GNSS, a bound system, is not affected by this putative explanation. However, this could be misleading, as the lack of impact on planets could be due to other characteristics, namely its much larger mass and composite nature, both making them poor "test" particles. If this is the case, then each individual satellite of the GNSS could experience anomalous effects: to ascertain these, one assumes a circular orbit, so that Eq. (5.7) reads

$$v^2 = \frac{GM_\oplus}{r} + \frac{\pi^2}{2}\beta c H r, \tag{5.10}$$

which, for a constant orbital radius r, induces a relative velocity shift

$$\frac{\delta v}{v} = \frac{\pi^2}{4}\frac{\beta c H r^2}{GM_\oplus} \sim 7 \times 10^{-10}, \tag{5.11}$$

so that the absolute velocity shift is of the order of 1 μ/s. This should give rise to an artificial change of the measured velocity of objects being tracked by the GNSS, with a similar order of magnitude: this is negligible when compared with the tracking precision required to probe the flyby anomaly, of the order of $\Delta v_\infty \sim 1\mathrm{mm/s}$ or less.

Conversely, if one assumes that the satellites are placed into orbits with a specified constant velocity, a shift in the orbit's radius of $7 \times 10^{-10}(R_\oplus + h) \sim 2$ cm is obtained, well below the typical spatial resolution discussed in the following sections. Thus, one concludes that the proposed difference between gravitational and inertial mass, if real, does not inhibit the use of the GNSS to yield a precise reconstitution of the trajectories of flybys.

5.3.3 *Modified Particle Dynamics*

Finally, owe examine how the modified particle dynamics model (discussed in Sect. 5.2.5) can affect the orbit of the global positioning satellites. Considering, for simplicity, perfectly circular orbits, one has $\mathbf{r} \cdot \mathbf{v} = 0$ (as in perigees), so that Eq. (5.8) reads

$$\mathbf{a}_A = A_{22}\frac{GM}{r^2}\frac{\mathbf{v}}{c}. \tag{5.12}$$

We assume that $A_{22} \sim 1$ as required to account for the flyby anomaly, considering $r = R_\oplus + h \sim 27{,}000$ km and taking the unperturbed value for the orbital velocity $v \approx \sqrt{GM/r} \sim 4000$ m/s, we get $a_A \approx 7 \times 10^{-6}$ m/s^2. Since this value is less than the typical 10^{-4} m/s^2 scale of the flyby anomaly, we can assume that any erroneous positioning arising from this effect does not affect its detection.

Also, the reader may notice that no increase of the orbital radius occurs, as one would expect in GR. Indeed, this would follow from energy conservation, which is not preserved in the model here scrutinised [6]. Strikingly, the same increase of kinetic energy that leads to the anomalous (positive) Δv_∞ in flybys is responsible for speeding up orbits while maintaining their radius.

Considering the discussion presented above, one may safely assume that the GNSS constellation would be fundamentally unaffected by the hypothetical flyby anomaly and may be thus employed to track probes performing gravity assists at the relevant region $h \sim 1000$ km.

5.4 GNSS Spacecraft Tracking

The tracking of spacecraft through GNSS systems is already commercially available (e.g. EADS-Astrium's Mosaic [26], NASA PiVoT [27]). These systems are typically used to follow satellites in Low Earth Orbit (LEO), at altitudes below those of the GNSS satellites ($h < h_{\text{GNSS}} \sim 20{,}000$ km), where the GNSS signal is stronger. Nevertheless, the Equator-S mission has shown it can receive front lobe signal from GPS satellites at an altitude of 61,000 km [28]. Furthermore, it is worth exploring the possibility of using the side and back lobes of the GPS signals to establish non-line of sight tracking and avoid the shading of the Earth [29, 30]. Clearly, the build up of more constellations and the use of multi-GNSS receivers, able to work simultaneously with different systems, will increase the accuracy of above-MEO satellite tracking in the near future.

The accuracy of GNSS spacecraft tracking is, understandably, better for lower orbits. However, it should be noted that during the apogee of Highly Elliptical Orbits (HEO), the velocity is, of course, much slower than close to perigee. This allows for the construction of a good orbital solution, despite the decreased signal coverage [31, 32]. As a result, the position and velocity accuracies for different types of orbit are somewhat similar, as outlined in Table 5.4 [31–34].

Table 5.4 Typical accuracies expected from GNSS satellite tracking systems for LEO, MEO, geosynchronous earth orbit (GEO) and HEO

Orbit	Apogee (km)	Position accuracy (m)	Velocity accuracy (mm/s)
LEO [33]	200–2000	10	10
MEO [33]	2000–GEO	30	20
GEO [33]	35,786	150	20
HEO [32]	>35,786	100	20

Recall that there is no full characterisation of the flyby anomalies and that these are detected from the mismatch between the expected and observed hyperbolic excess velocities after gravitational assist. As stated before, there is an inability of the DSN to track the spacecraft trajectories at low altitudes, close to the perigee, leading to a gap of approximately 4 h in the tracking data. Regarding the possibility of using the GNSS in this region, Fig. 5.2 shows that, although the velocity error is maximum close to perigee, this peak is very localised [32]. From a baseline of ~20 mm/s during the remaining orbit, it peaks briefly at ~100 mm/s during the first perigee approach, and converges towards ~50 mm/s in the subsequent perigee passes. By plotting the aforementioned gap, one sees that accuracies of ~20 mm/s are attainable during approximately half of this time interval.

For the study of the flyby anomaly, a high velocity accuracy would be of interest, at least of the same order of magnitude as the observed $\Delta v_\infty \sim 1$ mm/s. Currently available systems provide around 20 mm/s, which is clearly insufficient for such a study. However, the presented accuracies are related to real-time orbit solutions, which is unnecessary for the present objectives, and can undoubtedly be improved if offline processing is used, alongside other weak signal tracking strategies [31]. This, together with the increasing numbers of elements of the available and upcoming GNSS, lead us to conclude that it is indeed feasible to use the latter to test the flyby anomaly, if not with the current capability, then in the near future.

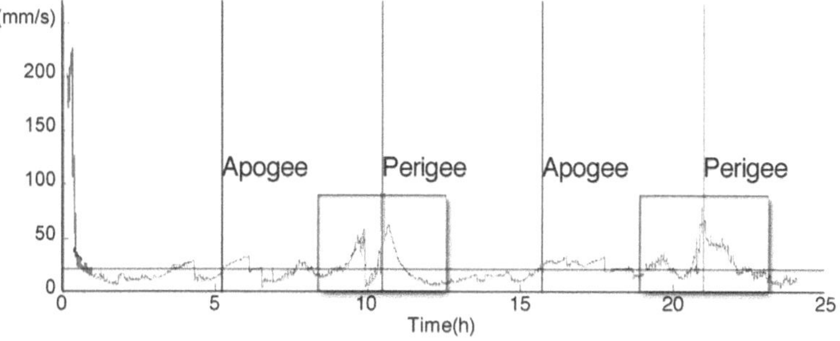

Fig. 5.2 Velocity error of multi-GNSS tracking of HEO spacecraft. *Boxes* (centered on perigee with 4 h width) signal the gap in DNS coverage; the *horizontal line* corresponds to a 20 mm/s accuracy. Adapted from Ref. [32]

5.5 Options for Probing the Flyby Anomaly

The issue currently at hand is still one of obtaining a full characterisation of the flyby anomaly, as its existence is not beyond dispute. Thus, regardless of the tentative physical models being put forward to account for it, one argues that GNSS tracking should be used to provide an independent method to study the behaviour of satellites in high-eccentricity low-perigee orbits independently from the traditional Doppler tracking methods.

It should be noted that the use of a constellation of redundant global positioning satellites enables one to discriminate between an anomaly affecting the motion of the flyby spacecraft and an anomaly in the propagation of light, which could be interpreted as the latter.

Firstly, any anomalous Doppler shift is disregarded, as time and position determination is obtained by resorting, not to the frequency of the transmitted signals, but to their propagation time.

Secondly, any anomaly in the latter can be detected by statistically analysing the set of individual transmissions. Closer GNSS satellites will have a smaller shift of propagation time, interpreted as a slight acceleration of the flyby spacecraft. Conversely, those farther away would indicate that the flyby probe has experienced a larger deviation. It should be noted that the typical height of the GNSS, $h \sim 20{,}000\,\text{km}$, is much larger than the considered perigees, of about 1000 km. Nevertheless, the difference between the closest distance possible at perigee ($\sim$19,000 km) and the largest one ($\sim$27,000 km), corresponding to a GNSS satellite in a plane perpendicular to the flyby trajectory) should allow for this statistical discrimination.

We consider two options to test the flyby anomaly:

(i) an add-on to an existing mission on a Highly Elliptic Orbit (HEO),
(ii) a dedicated low-cost mission in either HEO or a hyperbolic trajectory.

In the first option, the choice would be to piggyback a multi-GNSS receiver in an existing space mission. Since these receivers are relatively compact and with a small power consumption [26, 27], the host mission could be a small low-cost platform. At perigee, a highly elliptical trajectory would present a comparable (although smaller) velocity and height as the reported anomalous gravitational assists, with the added benefit of allowing for repeated experiments.

One can take as an example of a suitable mission the cancelled Inner Magnetosphere Explorer (IMEX) mission of the NASA University Explorer programme, with a mass of only 160 kg and a total budget of $15 M [35] in 2000. The IMEX probe was to be launched as a secondary payload on a Titan IV launcher, but was cancelled due to cost overrun. It would have followed a HEO, as summarised in Table 5.5, which would provide a "flyby" velocity at perigee of about 10 km/s, relatively close to the reported anomalous flybys.

The more ambitious option of a dedicated mission naturally has a number of advantages over the former, the main of which is the choice of orbit that can closely mimic a gravity assist, including an hyperbolic one. However, as discussed above,

Table 5.5 Orbital parameters of IMEX's highly elliptical orbit and similar hyperbolic flyby

	IMEX	Similar hyperbolic orbit
Perigee altitude	349 km	349 km
Apogee altitude	35,800 km	–
Velocity at perigee	10.1 km/s	11 km/s
Eccentricity	0.7248	1.04
Orbital period	10.5 h	–

a closed orbit of sufficiently high eccentricity would provide for multiple flybys, increasing the quality of the obtained data and allowing for a better characterisation of the anomaly. The HEO would also allow to ascertain if the flyby anomaly is exclusively linked to hyperbolic orbits. Also, possible error sources such as aerodynamic and thermal effects close to perigee could be more closely controlled with a dedicated mission. For instance, the spacecraft could be enclosed in a spherical radio-transparent body, so to simplify modelling and reduce directional effects. If a spin is given, any accidental anisotropies would be averaged out, yielding a much cleaner testbed for the desired experiment.

This mission would require a micro-satellite with a mass under 100 kg and a budget cap similar to the IMEX mission. This upper bound is rather straightforward to argue by comparison. Firstly, no additional spending is anticipated, due to the simplified spherical design over the more complex IMEX probe. Secondly, the scientific instrumentation found in the latter would be replaced by just a multi-GNSS receiver, thus lowering the total cost. More ambitiously, an added accelerometer could provide for a cost-effective independent measure of the acceleration profile, with a modest addition to the mass budget, such as the $\sim$3 kg μ STAR instrument considered in the Outer Solar System and Odyssey mission proposals [36, 37] or the accelerometer in the SAGAS project [38].

Following in this proposal, besides the feasibility of using GNSS to probe the flyby anomaly, the presented estimate illustrates the low cost of a dedicated mission for that purpose. Nevertheless, the actual cost could, in principle, be somewhat smaller than $15 M, the IMEX cost estimate, not only due to the inherently simpler design and instrumentation, but also due to the ongoing trend of decreasing micro-satellite costs, reflecting advances in miniaturisation, lower power consumption and improved industrial processes [39].

5.6 Conclusions

In this chapter, the use of the Galileo system or GNSS in general to study the flyby anomaly serves as the basis for a mission concept proposal.

We find that most of the available studies dealing with the tracking of spacecraft in real time have an insufficient velocity and position accuracy to detect this discrepancy. However, since this real time accuracy is only one order of magnitude above the required one (in particular, ~10 vs. ~1 mm/s in velocity), it is reasonable to expect that this situation could improve in the short-term. However, a more rational use of available resources could lead to a suitable tracking of spacecraft with greater accuracy, by abandoning real time solutions, and resorting instead to offline processing, use of side and back lobe tracking, amongst other weak signal tracking strategies. Crucially, the use of several GNSS at once should lead to an increased coverage of the different geometries.

Thus, it can be safely stated that there is no fundamental issue preventing the use of GNSS tracking to study the reported flyby anomaly. Naturally, this availability is not sufficient, as only spacecraft equipped with a (multi-)GNSS receiver would allow for such a study. In this work, we have shown that a mission of this kind could be easily deployed, either as an add-on package to an existing platform with the required highly elliptical orbit, or through a dedicated mission. While the first scenario would provide a cheaper solution, it is argued that a dedicated mission could be envisaged with a higher scientific payoff, while maintaining an overall low-cost approach.

Regardless of the actual origin of the flyby anomaly (unaccounted conventional effect, precision glitch or, more appealingly, new physics), this proposal offers a low-cost opportunity for displaying some of the scientific possibilities opened by the GNSS era.

References

1. P.G. Antreasian, J. Guinn, Investigations into the unexpected Delta-V increases during the Earth gravity assists of Galileo and NEAR, in *Proceedings of the AIAA/AAS Astrodynamics Specialist Conference and Exhibit, Boston,* 1998. American Institute of Aeronautics and Astronautics. doi:10.2514/6.1998-4287
2. J. Anderson, J. Campbell, J. Ekelund, J. Ellis, J. Jordan, Anomalous orbital-energy changes observed during spacecraft flybys of earth. Phys. Rev. Lett. **100**(9), 091102 (2008). doi:10.1103/PhysRevLett.100.091102
3. S.G. Turyshev, V. Toth, The puzzle of the flyby anomaly. Space Sci. Rev. **168**(1–4), 169–174 (2009). doi:10.1007/s11214-009-9571-0
4. Galileo Science Opportunity Document. Technical Report, European Space Agency (2010), http://egep.esa.int/egep_public/file/GSOD_v2_0.pdf
5. O. Bertolami, F. Francisco, P.J.S. Gil, J. Páramos, Testing the flyby anomaly with the GNSS constellation. Int. J. Mod. Phys. D **21**, 1250035 (2012). doi:10.1142/S0218271812500356
6. C. Lämmerzahl, O. Preuss, H. Dittus, *Is the Physics Within the Solar System Really Understood?*. Lasers, Clocks and Drag-Free Control (Springer, Berlin, 2008), pp. 75–101. doi:10.1007/978-3-540-34377-6_3. ISBN 978-3-540-34376-9
7. B. Rievers, C. Lämmerzahl, High precision thermal modeling of complex systems with application to the flyby and Pioneer anomaly. Ann. Phys. **523**(6), 439–449 (2011). doi:10.1002/andp.201100081
8. O. Bertolami, F. Francisco, P.J.S. Gil, J. Páramos, Thermal analysis of the Pioneer anomaly: a method to estimate radiative momentum transfer. Phys. Rev. D **78**(10), 103001 (2008a). doi:10.1103/PhysRevD.78.103001
9. F. Francisco, O. Bertolami, P.J.S. Gil, J. Páramos, Modelling the reflective thermal contribution to the acceleration of the Pioneer spacecraft. Phys. Lett. B **711**(5), 337–346 (2012). doi:10.1016/j.physletb.2012.04.034

10. N. Ashby, Relativity in the global positioning system. Living Rev. Relativ. **6**, 1 (2003). doi:10.12942/lrr-2003-1
11. C.W.F. Everitt, D.B. DeBra, B.W. Parkinson, J.P. Turneaure, J.W. Conklin, M.I. Heifetz, G.M. Keiser, A.S. Silbergleit, T. Holmes, J. Kolodziejczak, M. Al-Meshari, J.C. Mester, B. Muhlfelder, V.G. Solomonik, K. Stahl, P.W. Worden, W. Bencze, S. Buchman, B. Clarke, A. Al-Jadaan, H. Al-Jibreen, J. Li, J. A. Lipa, J.M. Lockhart, B. Al-Suwaidan, M. Taber, S. Wang, Gravity probe B: final results of a space experiment to test general relativity. Phys. Rev. Lett. **106**(22), 221101 (2011). doi:10.1103/PhysRevLett.106.221101
12. S. Adler, Can the flyby anomaly be attributed to earth-bound dark matter? Phys. Rev. D **79**(2), 023505 (2009). doi:10.1103/PhysRevD.79.023505
13. S. Adler, Modeling the flyby anomalies with dark matter scattering. Int. J. Mod. Phys. A **25**(24), 4577–4588 (2010). doi:10.1142/S0217751X10050706
14. S. Adler, Modeling the flyby anomalies with dark matter scattering: update with additional data and further predictions. Int. J. Mod. Phys. A **28**(15), 1350074 (2013). doi:10.1142/S0217751X13500747
15. R. Bernabei, P. Belli, F. Cappella, R. Cerulli, C.J. Dai, A d'Angelo, H.L. He, A. Incicchitti, H.H. Kuang, J.M. Ma, F. Montecchia, F. Nozzoli, D. Prosperi, X.D. Sheng, Z.P. Ye, First results from DAMA/LIBRA and the combined results with DAMA/NaI. Eur. Phys. J. C **56**(3), 333–355 (2008). doi:10.1140/epjc/s10052-008-0662-y
16. M.E. McCulloch, Can the flyby anomalies be explained by a modification of inertia. J. Brit. Interpla. Soc. **61**, 373–378 (2008)
17. M.E. McCulloch, Modelling the Pioneer anomaly as modified inertia. Mon. Not. R. Astron. Soc. **376**(1), 338–342 (2007). doi:10.1111/j.1365-2966.2007.11433.x
18. O. Bertolami, J. Páramos, S.G. Turyshev, *General Theory of Relativity: Will It Survive the Next Decade?*. Lasers, Clocks and Drag-Free Control (Springer, Berlin, 2008b), pp. 27–74. doi:10.1007/978-3-540-34377-6_2. ISBN 978-3-540-34376-9
19. O. Bertolami, J. Páramos, *The Experimental Status of Special and General Relativity*. Springer Handbook of Spacetime (Springer, Berlin, 2014), pp. 463–483
20. C.M. Will, The confrontation between general relativity and experiment. Living Rev. Relativ. **9** (2006). doi:10.12942/lrr-2006-3
21. B. Bertotti, L. Iess, P. Tortora, A test of general relativity using radio links with the Cassini spacecraft. Nature **425**(6956), 374–376 (2003). doi:10.1038/nature01997
22. O. Bertolami, F. Francisco, P.J.S. Gil, J. Páramos, Modeling the nongravitational acceleration during Cassini's gravitation experiments. Phys. Rev. D **90**(4), 042004 (2014). doi:10.1103/PhysRevD.90.042004
23. O. Bertolami J. Páramos, Using global positioning systems to test extensions of general relativity. Int. J. Mod. Phys. D **20**(09), 1617–1641 (2011). doi:10.1142/S0218271811019451
24. D.P. Rubincam, Drag on the LAGEOS satellite. J. Geophys. Res. **95**(B4), 4881–4886 (1990). doi:10.1029/JB095iB04p04881
25. R. Scharroo, K.F. Wakker, B.A.C. Ambrosius, R. Noomen, On the along-track acceleration of the LAGEOS satellite. J. Geophys. Res. **96**(B1), 729–740 1991. doi:10.1029/90JB02080
26. P. Krauss, Modernized spaceborne GNSS receivers, in *Proceedings of the 17th International Federation of Automated Control World Congress*, pp. 4707–4712 (2008). doi:10.3182/20080706-5-KR-1001.00792. ISBN 9783902661005
27. M.C. Moreau, P. Axelrad, J.L. Garrison, M. Wennersten, A.C. Long, Test results of the PiVoT receiver in high earth orbits using a GSS GPS simulator, in *Proceedings of the 14th International Technical Meeting of the Satellite Division of The Institute of Navigation, Salt Lake City*, pp. 2316–2326, September 2001
28. EQUATOR-S Home Page (2014), http://www2011.mpe.mpg.de/EQS/. Accessed 02 Nov 2014
29. M.C. Moreau, GPS Receiver Architecture for Autonomous Navigation in High Earth Orbits. Ph.D. Thesis, University of Colorado, 2001
30. A. Barrios-Montalvo, In-Orbit Autonomous Position Determination of Satellites Using Sparsely Distributed GNSS Measurements. Master's thesis, Luleå University of Technology, June 2010

31. M.C. Moreau, P. Axelrad, J.L. Garrison, A. Long, GPS receiver architecture and expected performance for autonomous navigation in high earth orbits. Navigation **47**(3), 191–204 (2000)
32. L. Qiao, S. Lim, C. Rizos, J. Liu, GNSS-based orbit determination for highly elliptical orbit satellites, in *Proceedings of the International Symposium on GPS/GNSS 2009, Jeju, Korea*, 2009
33. M. Mittnacht, E. Gottzein, M. Hartampf, J. Heim, P. Krauss, The mosaic GNSS receiver family, in *Proceedings of the 2nd ESA Workshop on Satellite Navigation User Equipment, Technologies, Noordwijk*, 2004
34. J.D. Kronman, Experience using GPS for orbit determination of a geosynchronous satellite, in *Proceedings of the 13th International Technical Meeting of the Satellite Division of the Institute of Navigation, Salt Lake City*, pp. 1622–1626, September 2000
35. IMEX-Inner Magnetosphere Explorer (2014), http://ham.space.umn.edu/spacephys/imex.html. Accessed 02 Nov 2014
36. B. Christophe, L.J. Spilker, J. Anderson, N. André, S.W. Asmar, J. Aurnou, D. Banfield, A. Barucci, O. Bertolami, R. Bingham, P. Brown, B. Cecconi, J.M. Courty, H. Dittus, L.N. Fletcher, B. Foulon, F. Francisco, P.J.S. Gil, K.H. Glassmeier, W. Grundy, C. Hansen, J. Helbert, R. Helled, H. Hussmann, B. Lamine, C. Lämmerzahl, L. Lamy, R. Lehoucq, B. Lenoir, A. Levy, G. Orton, J. Páramos, J. Poncy, F. Postberg, S.V. Progrebenko, K.R. Reh, S. Reynaud, C. Robert, E. Samain, J. Saur, K.M. Sayanagi, N. Schmitz, H. Selig, F. Sohl, T.R. Spilker, R. Srama, K. Stephan, P. Touboul, P. Wolf, OSS (Outer Solar System): a fundamental and planetary physics mission to Neptune, Triton and the Kuiper Belt. Exp. Astron. **34**(2), 203–242 (2012). doi:10.1007/s10686-012-9309-y
37. B. Christophe, P.H. Andersen, J. Anderson, S. Asmar, P. Bério, O. Bertolami, R. Bingham, F. Bondu, P. Bouyer, S. Bremer, J.M. Courty, H. Dittus, B. Foulon, P.J.S. Gil, U. Johann, J.F. Jordan, B. Kent, C. Lämmerzahl, A. Levy, G. Métris, O. Olsen, J. Páramos, J.D. Prestage, S.V. Progrebenko, E. Rasel, A. Rathke, S. Reynaud, B. Rievers, E. Samain, T.J. Sumner, S. Theil, P. Touboul, S.G. Turyshev, P. Vrancken, P. Wolf, N. Yu, Odyssey: a solar system mission. Exp. Astron. **23**(2), 529–547 (2008). doi:10.1007/s10686-008-9084-y
38. P. Wolf, Ch. J. Bordé, A. Clairon, L. Duchayne, A. Landragin, P. Lemonde, G. Santarelli, W. Ertmer, E. Rasel, F.S. Cataliotti, M. Inguscio, G.M. Tino, P. Gill, H. Klein, S. Reynaud, C. Salomon, E. Peik, O. Bertolami, P. Gil, J. Páramos, C. Jentsch, U. Johann, A. Rathke, P. Bouyer, L. Cacciapuoti, D. Izzo, P. De Natale, B. Christophe, P. Touboul, S.G. Turyshev, J. Anderson, M.E. Tobar, F. Schmidt-Kaler, J. Vigué, A.A. Madej, L. Marmet, M.C. Angonin, P. Delva, P. Tourrenc, G. Métris, H. Müller, R. Walsworth, Z.H. Lu, L.J. Wang, K. Bongs, A. Toncelli, M. Tonelli, H. Dittus, C. Lämmerzahl, G. Galzerano, P. Laporta, J. Laskar, A, Fienga, F, Roques, K, Sengstock, Quantum physics exploring gravity in the outer solar system: the SAGAS project. Exp. Astron. **23**(2), 651–687 (2008). doi:10.1007/s10686-008-9118-5
39. D.A. Bearden, Small-satellite costs. Crosslink **2**(1), 33–44 (2001)

Chapter 6
Conclusions and Outlook

In this thesis, we set out to explore three distinct, albeit related problems. Their connection lies in the fact that they all involve trajectory anomalies in interplanetary spacecraft and that all of them test the boundaries of General Relativity (GR). The work performed in connection with the Pioneer anomaly [1, 2], Cassini's gravitational experiments [3] and the flyby anomaly [4] led to, arguably, very significant contributions.

In the search for a solution to the Pioneer anomaly, an entirely new method was developed to address an issue that was also new: the reliable computation of thermally induced accelerations in spacecraft. The point-like source method was developed with transparency, flexibility and speed in mind. It was judged that a method with these characteristics was better suited to the problem at hand than the conventional approaches through finite-element models.

Crucial to the results of the Pioneer thermal modelling effort, was also the use of a parametric analysis through a Monte Carlo simulation to obtain a probability distribution for the final result, based on the more likely values of the input parameters. The results ultimately support the thesis of thermal effects being most likely responsible for the Pioneer anomaly.

As a consequence of the results presented in Chap. 2, the Pioneer anomaly was finally declared as a closed matter, after several years of controversy. Those who argued from the outset that it was a result of thermal effects, often against strong opposition, were ultimately vindicated.

Thermal effects were also the key to close the accounting of non-gravitational acceleration during Cassini's solar opposition experiment, designed to test GR to unprecedented accuracy. Decisive to this experiment is a tight control of non-gravitational accelerations, namely, those due to thermal radiation. The methodologies developed for the Pioneer allowed for a reliable computation of thermal acceleration, confirming the estimates obtained from radiometric data. The importance of this conclusion resides in a reinforced confidence on the gravitational parameters measured.

The undeniable success of the point-like source method in modelling thermal acceleration in spacecraft has been successively demonstrated by its ability to tackle

F. Francisco, *Trajectory Anomalies in Interplanetary Spacecraft*, Springer Theses,
DOI 10.1007/978-3-319-18980-2_6

the two problems it was presented with. The point-like source method is, unquestionably, one of the most significant outcomes of this thesis. Its success in obtaining reliable results for the thermally induced acceleration in spacecraft such as the Pioneers and Cassini is the best proof of its reliability and usefulness. The ability to deal with the uncertainties usually involved in the modelling of spacecraft launched years or decades before and exposed to the outer space environment is an asset that should still prove useful in future situations.

Contrasting with the results of the Pioneer and Cassini studies, many open questions surrounding the flyby anomaly still remain. While no convincing explanation has arisen, or even a full characterisation, the proposal discussed in Chap. 5 for a method to study the phenomenon is a starting point for future developments in this field. Still there are some barriers to overcome. While the unavailability of technology to track the spacecraft in real time near the perigee with sufficient accuracy and resolution can be surpassed with post-processing, there is still some theoretical work that needs to be made in order to discriminate effects specific to hyperbolic orbits.

A new generation of missions to test this theory with unprecedented accuracy is now being proposed. The OSS proposal, discussed in Chap. 4, continues on a tradition of combining fundamental physics experiments on planetary science missions, a tradition to which both Pioneer 10 and 11 and Cassini made significant contributions. The gravitational experiments included in the scientific objectives of OSS constitute an incremental step over the measurements made by the Pioneer and by Cassini.

Although two of the main problems addressed in this thesis are closed, there are some future lines of research spawning from this work.

The flyby anomaly remains unsolved and future opportunities for its study, namely, in future space missions must be seized. Also, we are still waiting for definite results on at least three recent flybys.

As to the overall theme of GR experiments, entirely new methods are now being put forward, for instance, in proposals such as GAME [5]. In order to make these new possibilities workable, there is still a considerable amount of theoretical work to be performed in order to identify which alternative theories can be discriminated and their observational signatures.

At least until then, General Relativity is set to complete one century and remain standing as one of the pillars of contemporary physics.

References

1. O. Bertolami, F. Francisco, P.J.S. Gil, J. Páramos, Estimating Radiative Momentum Transfer Through a Thermal Analysis of the Pioneer Anomaly. Space Sci. Rev. **151**(1–3), 75–91 (2010). doi:10.1007/s11214-009-9589-3
2. F. Francisco, O. Bertolami, P.J.S. Gil, J. Páramos, Modelling the reflective thermal contribution to the acceleration of the Pioneer spacecraft. Phys. Lett. B **711**(5), 337–346 (2012). doi:10.1016/j.physletb.2012.04.034

3. O. Bertolami, F. Francisco, P.J.S. Gil, J. Páramos, Modeling the nongravitational acceleration during Cassini's gravitation experiments. Phys. Rev. D **90**(4), 042004 (2014). doi:10.1103/PhysRevD.90.042004
4. O. Bertolami, F. Francisco, P.J.S. Gil, J. Páramos, Testing the flyby anomaly with the GNSS constellation. Int. J. Mod. Phys. D **21**, 1250035 (2012). doi:10.1142/S0218271812500356
5. M. Gai, A. Vecchiato, S. Ligori, A. Sozzetti, M.G. Lattanzi, Gravitation astrometric measurement experiment. Exp. Astron. **34**(2), 165–180 (2012). doi:10.1007/s10686-012-9304-3

Appendix A
Full Results for Point-Like Source Test Cases

Table A.1 Results for test case 1 (*cf.* Table 2.3 in Sect. 2.3.3) considering a total emission of 1 kW

Sources	Energy flux (W)	Δ (%)	Force components (x, y, z) $(10^{-7}$ N)	Force intensity $(10^{-7}$ N)	Δ (%)
1	15.34		(0.9300, 0, 0.1514)	1.004	
4	15.92	3.8	(1.028, 0, 0.1638)	1.041	3.6
16	16.09	1.0	(1.038, 0, 0.1675)	1.051	0.98
64	16.13	0.26	(1.040, 0, 0.1684)	1.054	0.25
144	16.14	0.049	(1.041, 0, 0.1686)	1.054	0.047

As the number of sources to represent the thermal emission of a surface change, the resultant force components appearing by shadow on the secondary surface remain almost the same

Table A.2 Same as Table A.1, for test case 2

Sources	Energy flux (W)	Δ (%)	Force components (x, y, z) $(10^{-7}$ N)	Force intensity $(10^{-7}$ N)	Δ (%)
1	19.20		(0.4952, 0, 1.037)	1.149	
4	19.83	3.3	(0.5032, 0, 1.082)	1.192	3.8
16	19.99	0.80	(0.5050, 0, 1.093)	1.204	0.92
64	20.03	0.20	(0.5054, 0, 1.096)	1.207	0.23
144	20.04	0.036	(0.5055, 0, 1.096)	1.207	0.042

F. Francisco, *Trajectory Anomalies in Interplanetary Spacecraft*, Springer Theses,
DOI 10.1007/978-3-319-18980-2

Table A.3 Same as Table A.1, for test case 3

Sources	Energy flux (W)	Δ (%)	Force components (x, y, z) (10^{-7} N)	Force intensity (10^{-7} N)	Δ (%)
1	26.13		(1.110, 0, 1.292)	1.703	
4	26.51	1.4	(1.111, 0, 1.320)	1.725	1.3
16	26.61	0.36	(1.111, 0, 1.327)	1.731	0.314
64	26.64	0.087	(1.111, 0, 1.329)	1.732	0.076
144	26.64	0.016	(1.111, 0, 1.329)	1.732	0.014

Table A.4 Same as Table A.1, for test case 4

Sources	Energy flux (W)	Δ (%)	Force components (x, y, z) (10^{-7} N)	Force intensity (10^{-7} N)	Δ (%)
1	44.41		(2.416, 0, 1.646)	2.923	
4	44.64	0.52	(2.409, 0, 1.663)	2.928	0.14
16	44.70	0.12	(2.407, 0, 1.668)	2.928	0.027
64	44.71	0.029	(2.407, 0, 1.669)	2.929	0.0059
144	44.71	0.0054	(2.406, 0, 1.669)	2.929	0.0011

Table A.5 Same as Table A.1, for test case 5

Sources	Energy flux (W)	Δ (%)	Force components (x, y, z) (10^{-7} N)	Force intensity (10^{-7} N)	Δ (%)
1	23.25		(1.395, 0, 0.4525)	1.467	
4	23.22	0.13	(1.383, 0, 0.4581)	1.457	0.68
16	23.20	0.067	(1.379, 0, 0.4593)	1.454	0.21
64	23.20	0.019	(1.378, 0, 0.4596)	1.453	0.055
144	23.20	0.0038	(1.378, 0, 0.4597)	1.453	0.010

Table A.6 Same as Table A.1, for test case 6

Sources	Energy flux (W)	Δ (%)	Force components (x, y, z) (10^{-7} N)	Force intensity (10^{-7} N)	Δ (%)
1	49.52		(0.6818, 0, 3.130)	3.203	
4	48.43	2.2	(0.6306, 0, 3.060)	3.124	2.4
16	48.16	0.56	(0.6190, 0, 3.042)	3.104	0.63
64	48.09	0.14	(0.6161, 0, 3.037)	3.099	0.16
144	48.07	0.026	(0.6156, 0, 3.036)	3.098	0.029

Table A.7 Same as Table A.1, for test case 7

Sources	Energy flux (W)	Δ (%)	Force components (x, y, z) $(10^{-7}$ N)	Force intensity $(10^{-7}$ N)	Δ (%)
1	50.36		(1.574, 0, 2.938)	3.333	
4	48.97	2.7	(1.501, 0, 2.855)	3.225	3.2
16	48.63	0.69	(1.484, 0, 2.834)	3.199	0.81
64	48.55	0.17	(1.480, 0, 2.829)	3.193	0.20
144	48.53	0.032	(1.479, 0, 2.828)	3.191	0.037

Table A.8 Same as Table A.1, for test case 8

Sources	Energy flux (W)	Δ (%)	Force components (x, y, z) $(10^{-7}$ N)	Force intensity $(10^{-7}$ N)	Δ (%)
1	45.53		(2.016, 0, 2.083)	2.899	
4	43.85	3.7	(1.918, 0, 2.003)	2.773	4.3
16	43.45	0.93	(1.895, 0, 1.984)	2.744	1.1
64	43.35	0.23	(1.890, 0, 1.979)	2.736	0.27
144	43.33	0.043	(1.889, 0, 1.978)	2.735	0.050

Table A.9 Same as Table A.1, for test case 9

Sources	Energy flux (W)	Δ (%)	Force components (x, y, z) $(10^{-7}$ N)	Force intensity $(10^{-7}$ N)	Δ (%)
1	24.29		(1.305, 0, 0.6316)	1.450	
4	23.47	3.3	(1.251, 0, 0.6113)	1.393	4.0
16	23.27	0.88	(1.238, 0, 0.6059)	1.378	1.0
64	23.21	0.22	(1.235, 0, 0.6045)	1.375	0.26
144	23.21	0.041	(1.234, 0, 0.6043)	1.374	0.048

Zeitfracht Medien GmbH
Ferdinand-Jühlke-Straße 7
99095 Erfurt, Deutschland
produktsicherheit@kolibri360.de